Applied Geography:

Practice, Problems and Prospects

Morgan Sant

Longman
London and New York

Longman Group Limited
Longman House
Burnt Mill, Harlow, Essex, UK

Published in the United States of America
by Longman Inc., New York

© Longman Group Limited 1982

First published 1982

British Library Cataloguing in Publication Data

Sant, Morgan
  Applied Geography.
  1. Geography
  I. Title
  910   G116   80-41371

  ISBN 0-582-30040-1
  ISBN 0-582-30041-X Pbk

Printed in Singapore by Huntsmen Offset Printing Pte Ltd

Contents

List of Figures

List of Tables

Preface

The seeds of this book were sown several years ago when, in the course of a late-night conversation with a friend (a geography graduate who had risen high in the planning profession) he asked the question: 'What can a geographer contribute to practical affairs?'

My answer was a rather conventional one: the education of well-rounded graduates, useful research, sensitivity to the many dimensions of environmental and regional issues, and so on. In truth I was not very satisfied with my answer, and neither was he.

Later I came to work at the University of New South Wales, where the School of Geography is administratively part of the Faculty of Applied Science. Although it offers courses in several faculties, the School has attempted to specialise in, and develop, programmes in *applied* geography, with separate streams in physical and economic geography. Most of the courses in these programmes are similar to those found in other universities, but there is a marked attempt to flavour them by focusing on career training. In being part of this development I have, in effect, been forced back to the earlier question: What can geographers working in an academic environment, contribute to practical affairs? Out of my searches for a better answer has come this book – which, I hasten to add, does not conclude the search.

Broadly, my answer is: a great deal, but it depends on their being aware of what is required of them when they become involved in praxis. It is not enough to expect others to value our efforts if we are not prepared to recognise their needs. (By 'others' I mean not only society in general but also those who would or could actively promote social welfare.) Mapping and measuring natural and man-made resources, designing ways of using them, describing and analysing the distribution of welfare, recognising and averting or avoiding hazards, framing and re-

viewing political and administrative institutions: these are some of the broad areas in which geographical applications can be of value, provided there is empathy between those working in the discipline and those outside it.

I know that not everyone will agree with all that is discussed here, or with the ideas that I have tried to express. (When the day comes that we all agree about everything the publishing industry will collapse.) Some will say that I have given insufficient space to techniques; some that I have not dealt adequately with political (especially radical) arguments; some that my concern for the integration of human and physical geography is misguided; some will be disappointed at the scant attention to project design; and some may feel aggrieved that I have overlooked their contribution to applied geography. To the last group I apologise.

What I have attempted is to capture what I perceive to be the essence of contemporary applied geography and to express what I believe should be included in its future. Inevitably, among the vast interests of geography (still the most fascinating of disciplines) there will be many that are not accorded proper justice in the confines of one person's specialisation and experiences.

I have concentrated upon what can be loosely called 'policy studies'. It ranges across a wide spectrum from the recognition of conditions that initiate policy or plan making, through the components of policy making, to its evaluation. Furthermore, I have focused primarily on *public* policy. The argument may be a little tenuous but I believe that the major questions of geography (concerning human and physical environmental *systems*, and their interaction) lead us to deal with populations rather than individuals and, hence, what most concerns us are social questions.

My approach to policy studies in the following chapters is thematic. Rather than work through various areas of public policy (e.g. housing, transport, land use, etc.) I have presented what I perceive to be major issues in the making of policy. Hence I have included chapters on information systems, evaluation, accounts and budgets, and so on. These may be unfamiliar terms but, on reflection, there has been much geographical work that contributes (sometimes indirectly or unintentionally) to these themes.

Obviously, while this book is personal statement it could not have been written without the many contributions made by friends and colleagues over the years, in seminars, discussions and arguments. They will recognise their influence (if it has not been dis-

torted too much) and there is no need to list them. If that seems like withholding names to protect the innocent, I have no such qualms about publicly thanking those who helped directly in producing my manuscript: Marion Cock and Diana Bryers who typed it, and Kevin Maynard and John Owen who prepared the illustrations. And, lastly, to Bridget, Christopher and Michael my thanks and love for their patience.

Morgan Sant
Paddington, NSW
May 1980

Acknowledgements

We are grateful to the following for permission to reproduce copyright material.

The Institute of British Geographers for a figure redrawn from W. Birch, *Trans. I.B.G.*, vol. 2, 1977; B.T. Batsford Ltd., for a table reprinted from R. Stamper, *Information in Business and Administrative Systems*, 1973; Pergamon Press, for extracts reprinted from T. Saaty, in Bernstein and Mellon, eds., *Selected Readings in Quantitative Urban Analysis*, 1978; Edward Arnold (Publishers) Ltd., for extract of a table in A. Young in Dawson and Doornkamp, eds., *Evaluating the Human Environment*, 1973; Longman Group Limited for extract of a figure redrawn from W.T.W. Morgan, *East Africa*, 1975; F.A.O., W.M.O. and U.N.E.S.C.O. for extract redrawn from the *World Map of Desertification*, 1977; Regional Studies Association for extracts of figures and tables redrawn from J. Forbes, *Regional Studies*, vol. 3, 1969; Lexington Books Ltd., for tables reprinted from S. Czamanski, *Regional and Interregional Social Accounting*, 1973; Little, Brown Inc., for a figure redrawn from A. Wildavsky, *Budgeting: a Comparative Theory of Budgetary Process-ess*, 1975; Oxford University Press for figures redrawn from J.B. Goddard, *Office Location in Urban and Regional Development*, 1975; McGraw Hill Ltd., for extract of a table in K. Smith, *Applied Climatology*, 1975; Oxford University Press for figures reprinted from I. Burton, R. Kates and G. White, *The Environment as Hazard*, 1978; Association of American Geographers for figures redrawn from P. Gould, *Annals, A.A.G.*, vol. 53, 1963; Regional Studies Association for figures redrawn from D.H. Green, *Regional Studies*, vol. 11, 1977; Oxford University Press for table and figure reprinted from B.E. Coates, R.L. Johnston and P.L. Knox, *Geography and Inequality*, 1977; Pergamon Press for figure and table reproduced from G. Chadwick, *A Systems View of Planning*, 1971; Hutchinson Ltd., for a figure redrawn from M. Roberts, *An Introduction to Town Planning Techniques*, 1974; Institute of British Geographers for tables and figure reprinted from W. Lever, *Trans. I.B.G.*, vol. 63, 1972; Edward Arnold (Publishers) Ltd., for a figure redrawn from D. Smith, *Human Geography, A Welfare Approach*, 1977; Doubleday Inc., for figures and extracts of tables redrawn from I. McHarg, *Design with Nature*, 1969; Water Resources Research for figures and tables redrawn from N. Dee et al, *Water Resources Research*, vol. 9, 1973; Yale University Press for a table reprinted from M. Deutsch, *The Resolution of Conflict*, 1973; Pergamon Press for figures redrawn from R. Morrill in Coppock and Sewell, eds., *Spatial Dimensions of Public Policy*, 1976; Lund University for figures which originally appeared in T. Hagerstrand, *Rapporter Och Notiser*, vol. 10, 1974, and T. Carlstein, *Rapporter Och Notiser*, vol. 11, 1974; Royal Meteorological Society for extracts of a table in J.B. Mason, *Quarterly Journal of the Royal Met. Soc.*, vol. 96, 1970; Liber/Gleerup (Publishers) for a figure originally published in A. Pred and G. Tornquist, *Systems of Cities and Information Flows*, 1973; Associated Book Publishers Ltd. for a figure redrawn from R.J. Bennett and R.J. Chorley, *Environmental Systems*, 1978.

1 Introspection

O Lord, give me the wisdom
To change the world today;
But let me use it
In a purely advisory way.

Lincoln Allison, 1975

The first part of this chapter presents, very briefly, a set of personal beliefs about the nature of applied geography. There is little attempt to justify these beliefs at this point, other than to put forward the statements of others who have also explored the same subject. Many of the points made here are elaborated in later chapters. The final part of the chapter contains a brief review of the formal place of applied geography within its parent discipline during the last two decades.

Ends and means

Applied geography is concerned about *ends* rather than *means*. It accepts axiomatically the view that the efficiency of our activities and the quality of our lives, now and in the future, are closely associated with how we use this planet and its resources, how we distribute ourselves on its surface, and how we relate to fellow men. But how we do these things depends upon decisions that we make, and how we exercise our choice from among the many alternative courses that we might follow. Often our decisions follow upon habit, rhythmically repeating the patterns of past choices. Sometimes they are innovative; and sometimes they are negative reactions to intended innovation. Applied geography, at its most basic, is the use of geographical knowledge as an aid to making those choices.

In this it is like other disciplines that have active and passive aspects, where knowledge is pursued not only for the enrichment of our understanding of physical and human milieux, but also as a means of adapting to, or changing, those milieux. Of course, we can justifiably expect there to be a symbiotic relationship between these two sides of geography but it would, I believe, be misleading to pretend that the two aspects do not exist as individual entities.

To write that the concern is with ends rather than with means may need some qualification. Naturally, there cannot be ends without means: and at a risk of being teleological, one end is only a means of moving towards another. In the present context, however, 'means' refers to theory and methodology. Neither of them is dispensable; decision making requires much more than intuition based on a reasonable education. Research – often a great deal of it – is generally a precondition for effective innovative action: and research cannot be undertaken without appropriate attention to theory and methodology. However, when we come down to detail there are different kinds of theory and methodology, just as there are different researchable questions. Not all of them assist in arriving at practicable conclusions. Most geographical research, with total justification, aims at deriving general statements of patterns or relationships in order to test hypotheses or to refine theories. Clearly we cannot automatically discount the relevance of this work to applied geography; 'there is nothing more practical than a good theory' (Bertrand Russell). But the link is often indirect, and the product of pure research generally cannot be translated immediately into practical action.

This distinction between 'pure' and 'applied' science is one of degree rather than kind, but it can be illustrated by reference to the way that each poses its questions. The critical difference is that applied science couches its questions in terms that are both *normative* and *prescriptive*, whereas in pure science the normative element may be present but not the prescriptive. From this there flow other differences.

To pursue a prescriptive course requires not only a need to specify questions in terms that relate closely (if not precisely) to some well-defined problem, but also that answers to those questions appear in a form that readily permits them to be subjected to further questions about the feasibility and desirability of the actions that they prescribe.

In substance, therefore, we would assert that pure and applied geography, though almost identical in subject matter, are distinguishable by what they aim to achieve and how they set about reaching their objectives. Nevertheless, they remain fundamentally one discipline. There are not two brands of geographers – pure and applied (or 'impure') – though some may have a predilection for one kind of work over another. There are, however, different modes of operation. In asserting this we differ significantly from Coppock, who wrote that 'the distinction between pure and applied research is not very appropriate in subjects such as geography' (1976). This view tends to ignore what may, in fact, be quite fundamental differences in motivation and *modus operandi*.

The history of applied science provides ample evidence that while science may be neutral its application, very often, is not. Indeed, why apply it, if not to reap some advantage, either personal or social? Our interest in this book is with the pursuit of long-term *social* benefit, which we see as being one of the two proper objectives of applied geography (the other being the advancement of the discipline itself). But we should be ready to admit that applied science can have cynical or malevolent motivations – and has had in the past. Thus Ratzel's conception of the organic nation-state, highly regarded by some (Dickinson 1969; Kasperson and Minghi 1970), was later distorted in the geopolitical writings of Haushofer in the 1930s to lend support to Nazi aggression (Weigert 1942). The practice of geopolitics continues to be intellectually disturbing (and morally bankrupt), but it should not be forgotten that some of its origins lie within geography even if it no longer receives much attention from geographers.

The lesson is not lost, however, for whenever we talk of *ends* we must also talk of *objectives* and the *values* that underlie these. The two most discussed motives for innovation – efficiency and equity – are concepts that pervade social science (Smith 1977). They are not the only objectives that men pursue, but without a clear understanding of their meaning and significance in any given situation the normative and prescriptive questions about that situation are almost certain to be deficient.

The scope of applied geography

Nowhere is the holism of intellectual activity more apparent than when we deal with applied science. For applied geography this works in two ways. Wilcock (1975), for example, warns against exclusivity: 'since geography is concerned, except in moments of rare aberration, to say something about the real world, it is unlikely that any worthwhile problem will be the exclusive preserve of trained geographers'. The corollary is partially stated by Coppock (1976): 'there is virtually no aspect of contemporary geography which is not affected to some degree by public policy'. He could, without contradiction, have included private actions as well as public ones. The consequence of this assertion, Coppock argues, is the need for geographers to identify their potential contributions to the resolution of policy issues and to gain greater access to policy makers.

Policy making and the monitoring of problems and plans have been at the heart of applied research. However, several geographers have been at pains to define quite strictly the role of the academic. Hare (1976) limits his function to 'aiding management and effectuation, rather than policy formation'. A similar view is expressed by Chisholm (1976) in a criticism of Blowers (1974) after the latter had contentiously claimed that 'geographers are among the best equipped intellectually to interpret social goals in terms of planning outcomes. . . . We already possess influence but that is not enough. What is needed now is the formulation of policies and the power to implement them.' The degree of engagement advocated by Blowers represents a view that few geographers would espouse, even though some have indeed been very close to, or actually involved in, the making of policy – including Chisholm himself. However, there are probably equally few who would concur with Stamp's distinction between 'the objective survey and analysis of the geographer . . . the subjective judgement of the planner' (1960). This carries disengagement beyond the limits that many would find desirable or realistic and, in any case, ignores what motivates much inquiry in the first place.

Behind any applied research there lies the sense of

a problem emanating from some observed or expected inadequacy of some defined situation. The in-
itiator of the study may be the researcher himself or
herself, but it is just as likely to be a body requiring
advice or assistance. But whatever the source, there
is, as Hagerstrand (1976) has pointed out, an important relationship between the academic and society;
'applied work on a broad scale in a social science discipline creates a very special kind of interaction between the society and that discipline'. There is a very
simple reason for this: namely that the geographer, like
other social scientists, must almost without exception
deal with communities or societies rather than with
individuals *per se*. (These comments apply equally to
physical geographers in their applied roles.) The onus
is therefore upon geographers to display the value of
their contribution to issues of *social* concern. Then,
assuming that they do this successfully, 'society' will
so order its affairs that the discipline will more readily be invited to contribute towards its planning and
policy making. This appears to have been achieved
to a significant degree in Sweden, due in no small
part to Hagerstrand himself. However, dissatisfac-
tion with the extent to which relationships have developed there – and barely at all in some countries –
prompted Hagerstrand to make the further observation that 'the main weakness of academic geography
from the applied point of view is that we have not
been able to handle the political dimension of human
affairs very well'. He is not very clear about the role
of the social scientist as social critic, although that
role is not inconsistent with the statements quoted
above.

A view of the scope of applied geography thus begins to take shape: the sense of problem, the contribution to decision making and policy, the monitoring of actions and the evaluation of plans. But
these are common to all applied social sciences. For
any single discipline to operate as an applied science
it must satisfy the condition that there are certain
social or environmental or technological conditions
that require (or could use) a contribution from it.

Again Hagerstrand provides a succinct definition
of the situations in which geographers would have a
direct interest: 'there must be locational shifts on a
substantial scale and also a willingness for planning
and control to guide change'. In this he is referring
chiefly to the location and allocation of people and
resources. But other situations are equally relevant:
for example, issues concerning the spatial organisation of activities or of institutions governing activi-

ties, in which 'locational shifts' of resources or
people are only indirectly affected.

The same definition applies to both physical and
human geography: indeed, the distinction between
physical and human geography is one that is often
difficult to sustain. Many human spatial problems
have major physical components: issues of environmental quality and land use are obvious examples.
Equally, the motivation for applied physical geography is either to help man adapt himself to his environment, or to indicate how the environment can
best be adapted to meet human demands. In either
case the location of resources or people, or both, is
affected (Hails 1977).

Who applies what?

Members of the academic profession are employed in
universities and colleges to teach and to conduct research, not to take decisions or to make policy about
matters outside their educational institution. They
may do so, but that is the outcome of a particular
agreement between the individuals and the bodies
using their expertise as advisers or consultants. In a
direct sense academics apply very little – which, of
course, makes a definition of 'applied' geography
very slippery, and it would be very easy to take the
permissive attitude that anything qualifies. But this
is not a particularly relevant issue. What is more important is the existence of channels whereby
academics can reach, and be reached by, those who
are capable of making or influencing decisions. Immediately, the sense of this statement must be made
clear. It is not to be taken as a belief in indiscrimin-
ate alignment with those controlling economic or
political power. For many, not necessarily radicals,
this would be academic prostitution. Rather, the
meaning is that applied geographers should be sensitive to problems, whatever their sources, and that
their concern should be in ensuring that the information that they have to offer is in a form that is useful
to, and usable by, those who need that information,
whether they are pressure groups looking for a more
equitable share of resources or government departments seeking more efficient operations.

Behind this is an acceptance of a division of
labour. Some might see this division idealistically,
with the academic cast in the role of a responsible

'fourth estate' – as social critic whose independence from the 'establishment' is a source of vitality. Although undoubtedly attractive, this is utopian. In reality the academic has a complex set of functions to perform. Apart from educating students who have then moved into a great variety of careers, geography has for many years been a primary source of graduates who have entered the professions of urban, regional and environmental planning. More recently there has been the emergence of formal training for planners, with their own university departments and institutions matching the growth of their profession. However, rather than reducing the demands on geography the reverse seems to have happened. The discipline is still the major source of entrants for postgraduate training and, equally important, it remains a major role of geography to conduct research that gives rise to innovation in planning concepts and methodology. Thus, through its continuing relationships with the planning profession it is impossible for geography to remain completely independent of the 'establishment'.

Of course, most planners do not 'apply' in a strict sense, either. Their role is to provide the intelligence for those who have the formal responsibility to make decisions (Chadwick 1971). However, for practical purposes the professional planner (and perhaps also the professional consultant) is often sufficiently close to the decision-making body that he may be only barely distinguishable from it, especially if the degree of direction upon his activities is relatively small and if there is a heavy reliance upon his advice.

In this situation the academic applied geographer is left with a difficult role unless he is content either to remain right outside the system or else to become so much a part of it that his work is determined by his 'clients'. First, his channels of contact are, at best, indirectly connected to those who take decisions and make policy, for the reasons discussed above. Second, the division of labour also implies that the burden of deciding what constitutes 'worthwhile' research lies largely with the academic. The choice of research may be institutionalised to some degree, by the influence of grant-allocating bodies (such as the research councils that exist in a number of countries), but generally the individual is free to conduct his inquiries how, and upon whatever topic, he pleases.

Given this lack of formal direction, the only guide to what is worth while is the academic's own observation of the present needs of society and of planners and decision makers, his anticipation of future needs (which means having an acute sense of timing) and the ability to select from the alternative topics that he could research, were he not constrained by time and resources. Given, also, that academics have beliefs, values and ideologies of their own, it is inevitable that a wide range of needs will be perceived and that much research that is motivated by the researcher's own sense of a problem will not be influential because it is not seen by others to be important or relevant. But that is a price of independence.

The role of theory

'Asking the right questions in the right way' has always been a precondition for successful research. However, as anyone who has ever tried to practise this principle will testify, there is no blueprint for achieving it. It is a far easier matter to recognise bad, incorrectly specified questions in a research proposal than it is to ensure that the best, or even the right, questions get asked.

However, a blueprint is not a guideline, and there are some principles that can be invoked when presenting hypotheses and research objectives (Amadeo and Golledge 1975). These are equally applicable to both pure and applied science since they share many of the same ground rules. However, they do diverge significantly in one respect. In applied science it is permissible (perhaps even required) to begin by posing the societal objectives to be achieved – creating so many new jobs in a region, or irrigating a certain number of hectares, or reducing flood risks by a given amount – and then applying knowledge already gained, or specially sought through new analyses and experiments, to 'solve' that problem. In pure science, on the other hand, the ends are unspecified: inquiry proceeds towards the creation of better theories, regardless of where the effort may lead.

Nevertheless, applied geography cannot stand in a theoretical vacuum and it is not merely concerned with problem solving, although that does provide a focus. Theory is essential at two levels. At one level, theory provides the framework for asking questions about the substantive relationships embodied in a problem. Issues such as demographic projections, the estimation of the return-period of severe floods, or forecasting the environmental impact of an oil refinery all require a theoretical basis to be conducted properly.

The alternative is intellectual anarchy, in which 'experts' can differ without there being an adequate basis for reconciling or accommodating their differences. Without some theoretical guidelines there would be no basis for selection from an infinite universe of possible predictions and prescriptions.

At the other level, theory becomes merged with philosophy to provide 'social theory' against which problems are identified as situations that are undesirable in terms of some moral or utilitarian goal or ideal. In the same vein, if a new goal or ideal emerges then things that previously were unexceptionable may take on great significance. Marxist geographers have been acutely aware of this level of theory and its role in shaping the nature of inquiry (Gregory 1978; Harvey 1973, 1974). So have many liberals, but in their case they have found it more difficult to formalise their beliefs in such an holistic normative manner.

Notwithstanding the importance of theory, there remains an acute problem in applied science of matching the variables that might be contained in a theory with the data that might be available to construct an operational model. The gap may be sufficient to call into question the very objectivity of the model if it has had to rely on proxy variables defined for convenience by the researcher. We shall return to this problem in Chapter 7 in a discussion of forecasting.

Space and place

Most modern geographical research has been conducted within a spatial paradigm, seeking spatial covariation between different phenomena. Its roots lie in the scientific positivism that has dominated the discipline during the last two decades. Within certain limitations this approach has been quite fruitful. For example, our understanding of how accessibility and mobility influence the distribution of resources and rewards, or the incidence of social goods and social bads, has been enhanced considerably. That the ability to generalise about spatial variations has been won at the expense of a concern for social and environmental detail and uniqueness – much of it vitally important – is often overlooked, but that is usually the outcome of one theoretical construct dominating over others. It inevitably imposes a methodology which stresses certain features while suppressing others.

Applied geography must be highly sensitive to this dilemma. On the one hand it must seek, and is the beneficiary of, theoretical advances that improve the understanding of what Birch (1977) has called 'the key elements of a coherent spectrum of geographical interest' – spatial differentiation, order, covariation, integration and change (Fig. 1.1) These advances provide tools for deductive reasoning. On the other hand, the *ultimate* interest of an applied geographer is generally not in space in general, but in a place or a set of places and communities. This restriction of interest – or greater specificity of problems – corresponds to the interests of decision makers who usually deal with particular resource allocation problems in particular places (and, we might add, at particular times). There is little chance of a policy being enacted and pursued if it is not directed at specific needs.

Sensitivity to the character and needs of places as well as a general understanding of systems of places is not only a recognition of reality. It also permits an enriched attitude to their planning, enhancing their individuality and internal variety. This notion is developed by Chadwick (1971) in a discussion of the 'law of requisite variety', in which it is stated that 'to control a system of given variety we must match it with a system of requisite (or similar) variety'. This means that high-variety systems, to be properly understood and 'controlled', require high-variety models. However, most spatial models are, in fact, low-variety and hence are not likely to be very effective. This, of course, is not an argument against spatial analysis, but it is a caveat against excessive expectations and potential misuse.

Quantification and evaluation

Around 1960 there was a vigorous debate about the desirability of quantitative geography: the quantifiers 'won' (Taylor 1976). The scientific method sought order and covariance and the way to do this was through measurement and statistical analysis (Burton 1963). To a large extent this is quite acceptable: numbers do matter. However, there are several statements that must be made to give a proper context for the role of quantification in applied geography.

Firstly, the purpose of much, perhaps most, quantification is to be able to make better qualitative judgements. The justification for this view comes

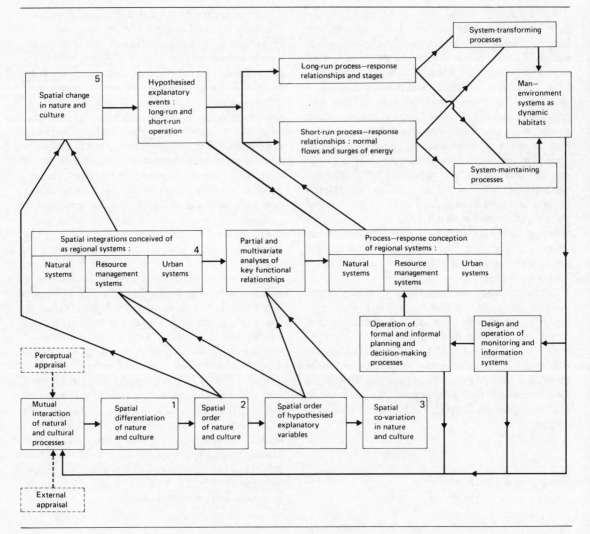

Fig. 1.1 A problem structure for geography (Birch 1977)

from the belief, shared by many economists, that utility can only be expressed in ordinal terms (Dasgupta and Pearce 1972; Mishan 1972) and that cardinal utility measurement is prevented by lack of knowledge and by uncertainty. When decisions are made, it is in the *expectation* that conditions will be improved or that welfare will be increased. There cannot be total confidence in the amount by which welfare will be increased.

Secondly, it is often impossible to divorce the things that we quantify from those that are readily quantifiable. This may not be a serious issue if the secondary sources are adequate or the problem is

sufficiently localised to permit good surveys. But at other times, especially when the researcher is investigating temporal change in complex regional systems, it is rare to find appropriate information for all the variables on all the dimensions that are needed. The corollary is that studies are often shaped by what is available. One only needs to witness the effect of the publication of a new census on the content of research in the next few years to see this.

Thirdly, as noted above, applied geography demands a degree of specificity accordant with the nature of decision making. The issues that concern decision makers *can* all be quite precisely defined: for example, poverty, accessibility, water quality, air pollution, and so on can all be described by simple indicators. More importantly, however, policies dealing with issues such as these need to be precisely and simply stated or they will be difficult to enact and implement. However, it is also possible, and may be desirable, to analyse these issues in very complex ways to identify their multi-dimensional nature. This may, if carried out clumsily, result in obfuscation, but the more important problem is that complex multivariate analyses may not be translatable into policy. The onus is upon the researcher to ensure that his results and conclusions really do have clear implications for specific actions if that is what he intends.

Fourthly, the numbers that policy makers are ultimately interested in often have monetary values attached to them. These are not always possible to calculate, as Lichfield, Kettle and Whitbread (1975) have cogently argued in their work on planning and evaluation; and, in any case, most geographers would find it difficult to conduct comprehensive economic analyses. Nevertheless, applied geography generally has to deal with real resources and quantitative analysis should be translatable to these needs.

Applied geography as information

What the preceding sections point towards is a role for applied geography as a supplier of information. Its purpose is to inform those who would, or could, either bring about geographical changes or determine the response to changing environmental conditions.

The information which geographers are, or could be, well qualified to impart comes out of the traditional interests of the discipline – human and physical environments and spatial relationships. Under these headings there occur issues too numerous to list individually but which fall under several broad headings: the use and conservation of natural resources; the structure and distribution of productive activities; the distribution of welfare. Each of these can be broken down into a long list of separate topics.

If this is a proper role for applied geography, then it is not undesirable that we should pay some attention to the nature and character of the information that we try to convey. This topic is discussed again in Chapter 7 in a slightly different context but the

Table 1.1 Classification of information according to its signification (Stamper 1973)

Signification	Mode	
Intention	Denotative	Affective
Descriptive	designations, facts, evidence, forecasts	appraisals, value judgements
Prescriptive	instructions, plans, policies, orders	inducements, coercion, threats, rewards

elements of that discussion also apply here. As applied geographers we are interested in both descriptive and prescriptive information, and also in both denotive (factual) and affective (judgemental) information (Table 1.1). We should, however, be aware of what kind of information we are dealing with at any one time.

In addition, we need to be aware that 'information' exists on a number of levels. Following Hagerstrand's comments (quoted above) we should distinguish, in Hermansen's terms (1969), 'the controlling system', or the political dimension, from the 'system to be controlled', or the social and physical phenomena in which we are interested. To make the study of the latter more practicable almost certainly requires a clear understanding of the former.

Thirdly, depending on the purpose of the information it is necessary to observe the characteristics that determine its quality: what information scientists refer to as syntactics, semantics and pragmatics (Stamper 1973). At the risk of oversimplification these can be defined as follows: the logical arrangement of data systems (syntactics); the meaning and interpretation of signs, symbols and words (semantics); and the social signification of statements (pragmatics). Arguably, we tend to pay greatest attention to the first and second. Since most academics spend most of their time dealing with other academics, this emphasis is neither surprising nor misplaced. Nevertheless, we can question whether sufficient attention is paid to pragmatic aspects, not only in terms of how geographers present their information but also – and more importantly – how information is used by others to bring about change. An example is provided by Blaikie (1975) in the diffusion of information relating to family-planning programmes in India, where the social context is obviously crucial to the effectiveness of the information.

Evolution

The term 'applied geography' is not new. It appeared in the late nineteenth century in the conferences of the International Geographical Union (IGU), where it described a section whose major interest was in what is now called medical geography. Official interest lapsed and the title was not revived until the IGU Congress of 1960. In the intervening years applied geography took on a wide meaning (Ackerman 1962). For example, two areas of interest were administrative regionalisation (Gilbert 1951) and land use survey (Stamp 1948). Later the Second World War saw geographers developing new applied skills in various information services. These included terrain analysis and air-photo interpretation (Taylor 1951). In the late 1940s and 1950s the growth of urban and regional planning was paralleled by a return to the earlier major interest – land use – with the emphasis shifting slightly from inventory to policy.

These developments are reflected in Sir Dudley Stamp's book, *Applied Geography*, first published in 1960, but partly anticipated in earlier writings (1957). This is a good testimonial to applied geography up to that time and, more important, it stimulated many young geographers in this aspect of their discipline.

The life and work of Dudley Stamp have been described in a memorial volume published by the Institute of British Geographers (Buchanan 1968; Wise 1968). Throughout most of his career, until his death in 1966, Stamp had played an influential role in the affairs both of his discipline and of his country. Undoubtedly his experience left him strongly placed to identify the use being made of his discipline by others and of the lacunae where there was potential use. It was this insight which had led him to initiate and organise the First Land Use Survey of Britain in the 1930s – a heroic information-gathering exercise that fitted well with Stamp's own view of planning, based on Geddes's trilogy: survey, analysis, plan. Stamp later remarked of that survey in 1960 that 'it had no ulterior motive . . . its objective was simply to record the factual position' (p. 39), a view that corresponded with his belief in the objectivity of the geographer being distinct from the subjectivity of the planner. Many present-day geographers would not make this distinction (Ley 1977).

The greater part of *Applied Geography*, a book that Stamp wrote towards the end of his career, reflects many of his own activities and interests, especially land classification and evaluation and land-use planning. It contains only a cursory review of urban, social and economic geography. Indeed the Land Use Survey, while giving detailed attention to agricultural land, treated urban areas cursorily and without differentiation. But it is not so much the substance as the spirit that is interesting.

In his discussion of the meaning and scope of applied geography, Stamp made several assertions that would now draw a mixed response. Firstly, the holism of geography was stressed; Stamp showed a traditional concern for the integration of man and environment. Possibly this reflected his abiding interest in land use, but he extended the same belief to his observations on population and industry. Perhaps not surprisingly this led him to be somewhat ambivalent on the subject of environmental determinism (a subject which only a few years previously had been the focus of vigorous debate culminating in the balanced argument presented by Montefiore and Williams (1955). Early in his book he states that 'we do better always to think of the influence of the environmental factors and of the reciprocal influence of man on those factors – even to their complete elimination in some cases' (p. 17). In contrast, towards the end, he asserts that '. . . the natural geographical factors are more important than they have ever been in the past. Man has not emancipated himself from the influence of those factors' (p. 194). Stamp was too pragmatic to be a root-and-branch determinist, but his early training in geology and his holistic approach to most issues tended to result in a less penetrating analysis of social and economic forces that most would now consider to be satisfactory. However, this was followed naturally by a call for interdisciplinary contact, an assertion that very few would contradict.

Secondly, Stamp displayed a profound respect for the power of maps: 'the geographic method of survey and analysis [is] achieved fully only when studied cartographically' (p. 9). In this regard modern geographers have proved to be highly ambivalent. Many have disavowed the use of maps (not only topographic ones, but thematic maps as well), seeing them as static representations of uninteresting facts. Yet at the same time a brief perusal of the academic journals will show that never before have maps been used so liberally or so adventurously, both as a means of expression and as a source of data. In some ways, there is no doubt, Stamp did exaggerate the value of maps, and a preoccupation with maps can divert attention away from more challenging process-oriented research. Nevertheless, in a discipline that focuses on the spatial dimension they remain the best means (along with meta-maps) of efficiently representing that dimension. They are also a vital component of geographic information systems, a topic to which we return in Chapter 7.

Thirdly, Stamp took a global view of events and conditions, particularly in his own field of expertise – land-use planning. His background must have contributed to this interest, having travelled widely and having played such a prominent part in the IGU (of which he was President during 1952–56). His own contribution – the formation of a World Land Use Survey – was not particularly successful, but it represents an important attempt by geographers to act internationally. 'Localism' is, of course, a regrettable fact of life for applied geographers. Most resource allocation problems are encountered within states, or even within regions or cities. But the larger scale still offers fruitful ground: the IGU commission on desertification and the UN Commission on the Environment represent two such forums for international research and co-operation.

Lastly, there is a deceptive innocence about Stamp's book which stems not from naivety but from confidence in his own judgement and experience. He had achieved much and his credentials commanded attention. Today we live in a less confident age. Perhaps this is because we have a greater propensity to invent complexities: perhaps the complexities are greater today. At any rate, the scope and methods of applied geography are more elaborate than they were a generation ago.

Applied geography and the IGU

In 1960, at Stockholm, there appeared for the first time in modern IGU congresses a section in the conference programme devoted to Applied Geography. There was also the adoption of a resolution inviting National Committees of Geography to present reports to the 1964 congress dealing with: (a) the form and content of applied geography and its relations to land development and land planning; (b) a directory of possible users of applied geography; and (c) the organisation of education and training for those wanting to take up careers in applied geography. Out of this activity there followed the creation, in 1964,

of an IGU Commission on Applied Geography, under the chairmanship of the Belgian geographer Omer Tulippe, who was succeeded by Michel Phlipponneau, who had published *Géographie et Action* in 1960.

The Commission's first programme, which occupied its first four years, followed the three points of the Stockholm resolution. Meetings were held in Prague (1965), Kingston, Rhode Island (1966), and Liège (1967) in order to exchange information and conduct discussions among geographers from many different countries. Following the 1968 congress in Delhi, the area of interest of the Commission was widened and subsequent meetings, at Warsaw (1970), Rennes (1971) and Waterloo, Ontario (1972), discussed such topics as environmental planning, urban morphology, new towns, regional planning and development, long-term forecasting and computer techniques. Another new programme was set out in 1972 at the Montreal congress, but by this time the momentum appeared to be waning and in the 1976 congress at Moscow the Commission was 'demoted' to the status of a Working Party.

The most obvious feature of all these deliberations is, unfortunately, a negative one. While most of the general discussion on the nature of applied geography and its educational content and role proved interesting and valuable, the substantive papers showed little cohesion and might just as easily have been delivered at some other commission or working party of the IGU. Taken together they showed little that was definitively 'applied'. Perhaps this was because of the weak conceptual underpinning of applied geography – a point that was made in the only major British contribution to the Commission, in which it was asserted that applied geography was pragmatic rather than conceptual (House 1966). In a comment at the Prague meeting, Freeman (1966) expressed his disenchantment in very strong terms: 'It may be that geographers, especially applied geographers, are too self-conscious and that it is most profitable to carry on with research work and then to see how it can be applied.'

In the event, little was said or done in the meetings of the Commission to dispel these serious misgivings. The conceptual basis of applied geography remained as undeveloped at the end as it had at the start. One regular contributor, Nash, was distinctive in his efforts to provide such a basis, associating applied geography with the creation of *future* geographies: 'Our challenges are no longer with the unmapped lands of the present or the landscapes of the past, but [with] the purposive delineation and investigation of the potential character and content of the future environment' (Nash, 1968). This was a brave attempt to provide a rationale, even though it was too narrowly focused towards forecasting.

However, all international gatherings are faced with severe constraints. There is a tendency to seek the lowest common denominator which, in this case, amounted to little more than an exchange of views. To gain a more realistic impression of the content and character of applied geography since 1960 it is necessary to look elsewhere than in the proceedings of the Commission.

2 Frameworks

The exercise of choice and the making of decisions *can* be simple processes, if the issues are familiar and straightforward. They can also be made to *appear* simple, if one is satisfied with hunches or prejudices. But then the risk of bad decisions is greater. On the other hand, complex or novel situations may pose profound difficulties and it is valuable, therefore, to have at least some general principles or guidelines to follow.

Hence the rationale of this chapter, in which we discuss points which may arise in a decision-making process, together with their implications for applied geography. It must be emphasised that there is no blueprint for correct decision making and no attempt is made here to present one. Thus the schematic diagram (Fig. 2.1) which is used to describe the points discussed in this chapter is no more than a convenient way of linking ideas together. Moreover, that diagram, and most of the discussion, focus on the public sector and are concerned particularly with economic and environmental planning at urban and regional scales. This is not to say that many of the same points do not also apply to private decision making, especially at the corporate level (Townroe 1971). If nothing else, both public and private decision makers have in common the description of planning attributed to Lindblom (1959) 'the science of muddling through'. But there are departures between public and private processes which make it important to deal with them separately.

In writing about applied geography it is not possible to avoid using the words 'decision' and 'decision making'. In one sense this is regrettable, for they convey an air of finality which is not always justifiable. Often a decision is only part of a sequence of stages in a long and, perhaps, convoluted route between a thought and an action. Imagine, for example, the stages leading to the purchase of a house: does one need a house? what sort of house? where? what price range? and so on. Also, what often

appears to be a final decision (say, one hallowed by Act of Parliament) is no more than a step in a long process of social conflict and experiment. So, if the words are unavoidable the meaning is not monolithic.

Frameworks for public policy

In a perfect world there would be a rational progression of steps between ideas and actions – or between the decision to take a decision and the implementation of the decision. This does not exclude the possibility of conflict and countervailing argument, but the processes would at least be conducted rationally. Although the real world is not like that (happily so, or it would be very dull), this ideal forms the basis of several modern texts on planning and systems theory (McLoughlin 1969; Chadwick 1971; Wilson 1974; Batty 1976; Bennett and Chorley 1978). Each of these authors employs a slightly different formalisation of public-sector decision making, but basically they are in agreement about the interrelations of the various steps: that is, goals – objectives – alternatives – evaluation – implementation – review, with the possibility of feedback loops between any pair of steps. They also agree about the value of such frameworks, not only as a means of arriving at decisions, but also as an aid to continuous monitoring. That is, if things appear to go 'wrong' it would be possible to backtrack along the process and identify the source of the problem.

Although the scheme in Fig. 2.1 reflects this thinking, we have to emphasise immediately that it is a gross oversimplification, for several reasons. Firstly, in reality, it would be usual to find that the steps occurred with confusing overlap; and that different people – often with conflicting motives and objec-

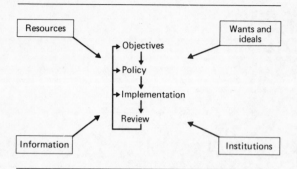

Fig. 2.1 Processes and influences in policy making

tives – would be responsible for different parts of the process. In short, the whole thing is a lot more messy than these frameworks suggest. Secondly, the apparent simplicity of the systems approach belies the difficulties faced in giving accurate and meaningful descriptions of its components. There are, for example, severe problems in identifying social objectives and evaluating alternative actions. Thirdly, while several of the components suggest that the systems approach is responsive to community wants and ideals (i.e. by stressing objectives and evaluation), it is easy for a system to be dominated by institutional frameworks and short-term economic vicissitudes. Hence the outcome might still be a system of bureaucratic processes and procedures (Gillingwater 1975). Fourthly, the systems approach makes no allowance for different modes of planning.

The last point is important because planning modes (the degree and nature of public intervention) influence the way in which different components in the systems approach are used. A useful typology, presented by McKay and Cox (1979) distinguishes between four modes. The first, *non-planning*, is self evident and has no significance for the systems approach. The second, *regulative* planning, is concerned solely with the limitation of initiatives coming from the private sector. The third, *indicative* planning, identifies the objectives towards which private and public resources should be guided, and uses controls and incentives towards this end. The fourth, *positive* planning, requires the state to control all resource allocation processes and to direct population and investment patterns towards centrally determined objectives.

Their distinctions may seem subtle, but there should be no doubt that the second and third of these categories place different demands on the operation of planning systems. Where one (regulative)

is orientated to the present, acting upon existing beliefs about the need to control socially undesirable actions, the other (indicative) is more orientated to future wants and ideals. Hence the latter must, by its nature, put more emphasis on forecasting, on goal-identification, on evaluation and on monitoring. In other words, it creates a richer climate for the blossoming of the systems approach to planning and policy making. And yet, many countries do not use indicative planning, and others fluctuate between indicative and regulative, or use one mode for certain areas of public involvement and another for different areas. Thus, in reality, the systems approaches described above are not universally relevant. Nevertheless, we shall persevere with the scheme outlined in Fig. 2.1 in the awareness that it is only a convenient way of arranging ideas and not a blueprint for decision making.

In this scheme there are two groups of components. The central 'core' contains the sequence of steps which might be followed in public policy making. Surrounding this are four items which can be referred to as elements in the decision-making environment. Thus the sequential core is surrounded by: (a) resources; (b) information systems; (c) institutions; and (d) wants and ideals. The first three are relatively simple concepts and need no discussion at this point: but the last item may suffer some ambiguity. Wants and ideals can have a number of different connotations. In this instance the words are used to express all contemporary values or aspirations. By not limiting their meaning, a distinction is drawn between them and the more radical (in the classical sense) term 'ideology'. They differ, also, from Gottman's (1973) notion of 'iconography', which can be interpreted to mean a *prevailing* ideology. Wants and ideals provide a wider context than either of these terms.

The framework contains two sets of linkages, firstly the four 'environmental elements' all relate to the central core in a reciprocal manner. Each one is a prerequisite for policy making and each may also be the subject of policy making. This merely recognises that policy has to be about something and cannot take place in a vacuum. Perhaps the link with resources (their creation, use or distribution) is the most familiar one in this context, but it is by no means uncommon to find policy towards the structure of institutions; or towards the quality of information systems; or to give status to the expression of aspirations (for example, by establishing and formalising public participation in planning). Secondly, each of the four elements relates to all of the others. Wants and ideals are formed about resources; institutions inspire and shape ideologies; resources influence the dimensions and quality of information systems (and, perhaps, the quality of their use); and so on.

The core sequence in public policy making

The quality of a decision may not be dependent upon the formality of the stages that preceded it, but there can be little doubt that it helps to know: (a) why a decision is being made (its goals and objectives); (b) what are the alternative policies and policy instruments that might be used in pursuing those goals and objectives; (c) how effective the alternative actions are likely to be; and (d) what the effects are likely to be upon other policies. Even with such a background, there is no guarantee that decision making will not ultimately degenerate into a lottery.

Unfortunately, this often appears to have been the case in much public policy. Examples abound of how agencies found themselves having to evaluate situations that had arisen because initial reviews had been insufficiently critical, not only of the potential costs and benefits of schemes but also of their overall resource requirements and their potential bottlenecks. One could go further and assert that more attention should have been paid to the opportunity costs of some major schemes. Two such examples have been reviewed by Brooks (1974, 1976), namely the Third London Airport, and the development of the supersonic aircraft, Concorde. In the first of these, severe criticism can be made of the failure to provide the research team with terms of reference that included a review of the need for another airport in the first place. In the absence of such terms of reference the results could only be in the form of comparative costs of the four alternative sites. In the second case, the decision to move to the research and development stage of a supersonic airliner and then to its actual construction was shrouded under several motives – none of which had much to do with commercial practice. The wish to exploit advanced technology; the need to show international co-operation (with France) prior to negotiating entry to the Common Market, the nationalistic (or strategic) demand to maintain a viable aircraft industry, and (having got the project well on its way) the wish to avoid unemployment in the Bristol region: all of these may have played a part. Whatever the principal reasons, the

outcome was a rather supine acceptance of wildly escalating costs.

Thus while it may not, in practice, be possible to achieve a coherent and effective route through the elements of the core sequence, it is still worth keeping it as a conceptual map to assist in reviewing and making decisions. Two points are particularly important. One is that each of the core elements in Fig. 2.1 serves the element immediately above it. Thus, means of implementation are defined and limited by the selected policy or strategy which, in turn, is designed to meet the objectives initially laid down. The other point is that each of the elements is likely to contain several choices. Thus, one objective may be equally well served by several policies, though not all of them will necessarily be equally easy to implement or incur the same costs.

An example may help to clarify these issues. In most countries that have regional policies the *objective* is to create greater equality between richer and poorer regions, or between regions of higher and lower unemployment. This objective has usually been stated in ordinal terms. Since the problem has tended to be seen as inequality in demand for manpower, rather than simply an excess supply of labour in poor regions (Richardson 1978), the ensuing *policy* has been to create employment in poorer regions, or to divert it there from richer regions. Among the *instruments* designed to serve this policy, areas have been designated for assistance, incentives and subsidies have been defined and, in some cases, controls imposed on the growth of richer regions. Finally, particular agencies have been charged with the responsibility for *implementing* and administering the policy.

This is a formal structure of regional policy that became familiar in many countries in the 1960s (Cameron 1970; Allen and McLennan 1970; Brewis 1969; McCrone 1969; Berentsen 1978).

In the British case it was possible to subject the process to a degree of evaluation not carried out elsewhere (Sant 1975). The results were illuminating, for it was clear that while the instruments were highly effective in serving the policy, it was also highly debatable whether the policy of relocating manufacturing employment achieved the objective of closing the gaps between regions in unemployment and earned income. The political realities demanded a regional policy, but more thought appeared to have gone into refining the instruments than into asking whether the objectives were achievable. A great deal of employment was created but the total was a long way short of sufficient to bring parity between regions. Some numbers calculated by Ridley (1972; see Sant 1975) illustrate the problem. Using a set of 'objectives' defined by himself but derived from government statements, Ridley estimated the number of new jobs required in the assisted areas to bring approximate parity in labour demand with the remainder of Britain. The answer – about 1 million jobs over a period of only ten years – was about double the number created directly by regional policies during 1945–66, and represented an annual rate of job creation substantially higher than in 1966–71 when the policy was pursued vigorously.

Objectives

In the 1960s a strong movement developed among professional planners for what came to be known as 'management by objectives'. Later there was equally strong criticism of this fashion. The initial concept was that if one could clearly and effectively define the objectives of a project or a plan then one was well on the way towards successfully implementing it. On the surface this is not unacceptable. However, where it can go wrong – and what prompted the criticism – is in the strong chance that not all relevant objectives will be defined and that those which are included will not be appropriately weighted. There was also the feeling that other things would be forced into a shape determined by a planner's own perception of objectives. Thus, if we assert that the defining of objectives is still a key to successful planning, it is in the light of the above comments. That is, objectives need to be defined exhaustively and they need to be treated sensitively.

This is done, firstly, by recognising where objectives originate: namely, among the wants and ideals of everyone who is affected by a project or a plan. For example, in most projects one can identify several broad groups: those who are direct recipients of the benefit and those who directly incur the costs; those who are responsible for administering the project; and those who are affected indirectly or incur external costs or benefits. Each group has a stake in what is done, which can be expressed in its own wants and ideals, and these can be expressed in its objectives. One group wants an expensive project with major benefits to itself; another group wants a cheap project; another doesn't mind, so long as the project is located far away; and so on. Secondly, exhaustiveness and sensitivity are more likely to be achieved by not discarding objectives (and thereby, not ignoring people's wants or ideals) at any stage in

the plan-making process until the final moment of plan selection. Some would argue, not even then: if people suffer from a project they should be compensated. Thirdly, it means weighting objectives equitably rather than simply assuming that each one is equally important. Some would argue that this happens automatically when a project is subjected to cost–benefit analysis; that is, the strength of wants and ideals is found in people's willingness to pay for benefits in comparison to the resource costs of a project. Others have argued that cash values provide inadequate weights (Lichfield *et al.* 1975; Hill 1968, 1973). These issues are discussed at greater length in Chapter 8.

Although what has just been described appears to be unwieldy, it is broadly in line with current practice. However, where practice deviates from concept is in its handing of objectives. Firstly, one cannot continue, *ad infinitum*, in the quest for objectives, and most plans appear to have concentrated on the wants and ideals of major groups of the population affected, together with objectives derived from 'official sources'. Secondly, objectives tend to have been classified by planners in ways which are designed to make them easier to handle. For example, the West Midland Regional Study Group (1971) established hierarchies of objectives, passing from general aims down to specific actions to meet those aims. In another plan (Coventry City Council 1971), objectives were divided into *essential* and *discriminatory* categories. A third example, the Strategy Plan for the North West (Department of the Environment 1975), dealt with *operational* and *regional* objectives. The former category consisted of bureaucratic and administrative needs – feasibility, ease of implementation and efficiency. The latter concerned environmental, economic and social conditions.

Clearly it is possible to draw up a list in which some objectives conflict with others. That is, one recognises the variety of people, and their wants, who will be affected by a plan or project. The task of identifying potential conflicts can be assisted by using a simple device, a goal-compatability matrix, as described in Fig. 7.5.

Lastly, there is a problem of quantification. On the surface it might appear that the planning process is made much easier when cardinal numbers can be incorporated in the objectives. However, this requires a degree of certainty that is sometimes difficult to attain, and which may not, in fact, be necessary. What is more important is the ability to register progress towards meeting objectives. In other cases, quantified objectives may be obtainable from forecasts and statutory requirements. For example, a forecast that there will be a certain number of school-age children in an area will be reflected in an objective to supply the appropriate number of school places.

Policies and strategies

When wants and ideals have been given formal expression it is then necessary to ask what processes, related to the distribution, allocation and development of resources, could be pursued in order to meet the objectives. These processes take their formal expression as policies or strategies (we use the words interchangeably).

At this point one moves into an intellectual domain where fact and theory are closely interrelated. That is, having specified an objective, we need: (a) to examine what this means 'on the ground', in the context of existing distributions and allocations of resources; and (b) to develop a hypothesis about actions that might meet that objective, given the real-world context. From the juxtaposition of fact and theory may come a policy or strategy.

A simple hypothetical example may illustrate this. A plan is to be created to deal with anticipated rapid population growth in a region. One objective is to avoid additional congestion on the region's road system without undertaking new, expensive investments. The existing structure of the region shows a dominant centre containing most of the region's office and service employment. Theory tells us that congestion is a function of the distribution of land uses: segregation of residence and workplace and a radial system focused on a single centre are likely to exacerbate congestion. Out of all this, one may derive a strategy to accommodate the extra population, perhaps in the form of a 'self-contained' new settlement with a mixture of residential and employment opportunities. But this is only one possible objective and strategy. The same problem might inspire another objective: for example, to minimise the consumption of rural land in residential and industrial development. This might prompt a different strategy: namely, to accommodate growth in, and on the fringes of, urban areas.

In essence, therefore, the preparation of a plan may require the testing of a number of different, perhaps conflicting, strategies to discover one that is preferred. Each alternative is likely to possess different characteristics and to give rise to different impacts on the distribution and consumption of resources.

An important question in testing alternatives is to assess the extent to which a strategy amplifies or, conversely, deviates from existing trends. This distinction has significance for estimating the feasibility of strategies: one that is *trend-amplifying* is likely to be easier (politically and economically) to implement than one that is *trend-deviating*. That is why, for example, transport and land-use plans which have been built on an acceptance of rising private car-ownership rates have been easier to formulate and implement than plans which have tried to promote public transport. This may not always be the case: if fuel costs make private motoring less attractive then public systems might become trend-amplifying.

The distinction, we should emphasise, relates to ease of implementation rather than to need or social benefit. A trend may only reflect private wants expressed through an imperfect market system. And, as is well known, private actions may impose social costs which outweigh private benefits. Hence a trend-deviating policy may be the preferred one. This has been argued in many instances related, for example, to traffic management (Starkie and Johnson 1975), environmental quality (Hardin 1968) and regional economic policy (Moore and Rhodes 1973). Of course, the ideal situation would be where a trend-amplifying strategy maximised net social benefits. To some extent this was sought in French regional policy through the designation of *métropoles d'équilibres* – the eight largest provincial urban systems – to act as counterweights to the attraction of Paris. As large centres they already possessed many of the conditions for growth, and the policy was to reinforce their advantages through public-resource allocation (Allen and McLennan 1970). In contrast, a 'worst-first' policy, trying to rescue regions in severe decline, starts from a position of much greater difficulty (Cameron 1970).

When faced with alternative policies that could be followed, one needs to be able to evaluate each one effectively so that a preferred policy can be identified. Evaluation can take a number of forms (as discussed in Chapter 8) which can broadly be put into two categories. One deals with feasibility and concerns the ability of the system actually to implement the policy. This needs certain questions to be resolved: is the bureaucratic structure appropriate for the policy? what other policies might be affected if this one was implemented? are there sufficient public resources? if private resources are to be mobilised, are they sufficient in quantity and flexibility? The other category deals with social costs and benefits and concerns such questions as: the contribution of the policy to aggregate welfare (i.e. its *efficiency* in resource use); its distributional effects, or who gains benefit and who incurs costs; the degree of uncertainty surrounding the effects of the policy; and whether the effects of the policy are short- or long-term.

Instruments and implementation

From a bureaucratic point of view these are often the most important elements in the policy-making sequence. This is not surprising, since at this point the questions arise: *who* should implement a policy? within what statutory or legal frameworks? and using what budgetary procedures? In a simple, single-sector plan these questions may not arise; elsewhere their resolution may profoundly influence the eventual shape of a policy. It is then appropriate that they should be closely scrutinised: for example, if a policy required a major shift in balance between local and central government it would be quite right to question the legitimacy of the policy. However, where the problem could be reduced to mere interdepartmental rivalry this could be regarded as a case of the 'tail wagging the dog'.

The design of policy instruments, therefore, is a task which may involve a number of practical issues. Firstly, it requires an assessment of the degree to which existing instruments can be adopted to serve the policy and whether it is necessary to devise new instruments. For example, in many countries if a new housing policy was proclaimed it would be possible to accommodate it by expanding existing agencies responsible for finance and construction. In contrast, a similar but more specific programme for constructing quite large new towns (say, to house a quarter of a million people) would probably be: (a) beyond the scope of any single local government to handle; and (b) a small proportion of a central government department's responsibilities, and thus may not get all the attention it needed. Hence, a special authority may be required.

Secondly, policies may be pursued in different ways. Instruments may take the form of *controls* or *incentives*, or both simultaneously. In general, a regulative planning mode tends to rely on controls in order to prevent unwanted actions taking place at particular locations, but otherwise leaving people to do as they please. Most land-use planning falls into this category. But regulative planning tends not to be creative: it only prevents certain activities. Short of using positive planning, creativity requires incentives

which are large enough to induce desired actions. The clearest example of this is found in regional economic planning, which can rarely be carried out fully unless the private sector co-operates. However, it is possible to design a great number of combinations of controls and incentives, and also to 'experiment' with them over time.

Again regional policy provides an example (McCrone 1969). In Britain since the 1930s the instruments used to promote the creation of jobs in the assisted areas have included investment incentives on buildings and machinery, tax incentives, and labour subsidies, all at different rates. At the same time there have been controls, first on manufacturing investment and later on office development in the more prosperous regions. These, too, have worked with different stringency at different times. From time to time there have also been changes in the boundaries of assisted areas. Added to these, central government has provided industrial estates and used the location of publicly owned industries to reinforce the policy, and local government has also played a part by promoting and providing for industrial development (Camina 1974). When, in 1972 the government announced that it would keep the system stable this was greeted with relief by the main employers' association, not only because it was happy with the level of subsidies but also because the pace of change in policy instruments had been bewildering. Perhaps in this case there had been too many experiments, making it difficult to evaluate the effects of each one. But attempts have been made (Sant 1975; Brown 1972; Moore and Rhodes 1973, 1974) which point to the greater effectiveness of certain instruments – advance factories, labour subsidies and locational controls – rather than others in getting firms to locate or create jobs in assisted areas.

Thirdly, there is a question of whether to use direct or indirect measures, which only partly overlaps with the preceding question of controls and incentives (i.e. controls tend to be direct; incentives can be either). To a certain extent the decision is a function of what one perceives the problem to be. For example, the issue of accessibility to services in a rural area might be seen to be a personal problem (certain people are too poor to 'buy' a sufficient level of mobility) or it may be seen as a broader issue of settlement structures and service location patterns. In the first instance the appropriate action might be to directly subsidise personal travel: in the second, it might be to take the longer-term, indirect, approach of changing the location of residences and services

(Moseley 1979). Similar issues arise in environmental planning where the choice may be between affecting individual sites and users or acting indirectly by imposing general standards on everyone. The problem in both cases is that indirect (and general) instruments tend to treat people and places equally but inequitably. A general environmental policy, for example, might inflict penalties on the 'innocent' and give unsought benefits to others or create a 'free rider' problem (Dales 1968). However, it may be the case that this is the only practicable way to operate, since a direct instrument would require a level of fact-finding about individual cases (people or firms) that might be beyond the capacity of even a large bureaucracy.

Environmental elements of the public-policy framework

The core sequence of policy making discussed in the last section does not exist in isolation and, indeed, could not be discussed without some reference to the four environmental elements depicted in Fig. 2.1. Moreover, while it is important as a coherent guide it is, in the end, no more than a schematic version of a possible procedure. It has no substantive status: it is not what policy is *about*. Nor does it take into account the constraints and opportunities imposed or offered by the environmental elements. These issues are taken up in the present section, which looks in turn at wants and ideals, resources, information systems and institutions. The discussion is brief since each of these gives the focus of chapters to follow.

Wants and ideals
These might properly be considered as the prime movers of the planning process in particular, and of changes in human landscapes in general. While what eventually is implemented may be what is 'expedient' or 'pragmatic', it is a truism to state that little, if anything, would be implemented if there were no ideals or wants.

Concepts of ideal political states, ideal cities and ideal environments have long been closely related. Aristotle's ideal state was, in fact, the city-state (Kasperson and Minghi 1970). Since the Renaissance the same three sets of ideals have been expressed in a stream of utopian thought and practice (Benevolo 1971) passing through Thomas More and Francis Bacon to the nineteenth-century utopian experimentalists, like Robert Owen in Britain, le Char-

din in France, and the New Harmony Society in the United States, to Ebenezer Howard and the New Towns Movement in the early twentieth century, and, more recently to the 'commune' escapism of the present. Utopia is a construct of the imagination – both 'no place and a good place' (Goodey 1970), and with rare exceptions utopian ideals have never been successfully implemented: in many cases they were not meant to be. What they express are *principles* of human organisation and behaviour which transcend what their authors saw to be the normal utilitarian behaviour of contemporary society. Moreover, as Porter and Lukerman (1976) describe, there has been a variety of utopian conceptions, consistent with the different character of their designers. Two major 'types' which they recognise in American urban development are the 'autarkic and natural Eden' and the 'geometric and planned New Jerusalem'. Modern town planning tends to show the same dichotomy (Banham 1969).

These are exaggerated expressions of idealism though they serve to draw the contrast with utilitarian philosophies. On a practical level, however, we can regard as idealistic any argument which promotes a greater level of consumption of a good than would occur if people were left to control their expenditures and activities entirely by themselves: or, conversely, a smaller amount of an 'undesirable' good. This raises the concept of *merit goods* which is discussed in the next chapter. A commonly cited example of a merit good is education, where the cost of schooling and the loss of earnings of the teenager might combine to reduce school attendance. The response in most national education systems has been to institute minimum leaving ages and subsidies to students at all levels, including the tertiary. But there are many other examples which have been promoted with greater or lesser degrees of acceptance: clean air and water; cultural facilities; health facilities; recreation facilities; fine architecture; beautiful landscapes; housing; and so on. When any one of them is promoted as a merit good it will almost certainly also be argued that it should be treated as a *public good* as well – that it should be provided from public revenue and that no member of the community can be prevented from enjoying its benefits. Even if this does not happen, a case will be made for public intervention in private patterns of expenditure. This is the essence of the 'welfare geography' about which David Smith (1977) has written eloquently and at length.

However, it is not enough to look at ideals alone:

the reason why something becomes a merit good (or is argued to be one) has its origins as much in the existing distribution of resources and, hence, in effective demand, as in some abstract concept of a 'good society'. Furthermore, the debate about merit goods extends beyond the issue of any particular good. At the very least it is a conflict over the importance of traditional economic goals (and, indeed over the role of economics as a discipline, as Galbraith (1975) has cogently argued). At its most simple, the issue is not one of production *per se*, but of what to produce and how to distribute the rewards according to other criteria than utilitarian ones. Rostow (1971) sees this as being a characteristic of nations that have passed into the stage of high mass-consumption: 'The shift from high mass-consumption to the search for quality is a real discontinuity in the life of man in industrial society – in one sense the first major discontinuity since take-off.' Despite the conscious hyperbole Rostow captures the mood of contemporary politics, not only in the United States, which is Rostow's main focus, but also in other rich countries.

There is, therefore, a clear and strong distinction between wants and ideals. But it would be quite false to regard public policy making as being motivated by ideals alone – or even, necessarily, in the main. Utilitarian wants, whether privately or publicly organised, are at least equally powerful sources of policy making. Moreover, unlike ideals, which rely on the appeal to some 'higher' aspirations and the force of principle, wants have the backing of personal or group interests, focused on the acquisition or defence of real resources. Not surprisingly, therefore, *interest* groups are as much concerned with the formation (or rejection) of policies as are *principle* groups. It is also not surprising to find the two groups in conflict with each other (or, for that matter, to find a conflict between two groups in the same category).

It is not always necessary for a want or an ideal to have strong political salience before it can be incorporated into the policy-making process. Examples, such as disaster relief (not unimportant as hazards appear to be increasing through man's own activities according to Burton, Kates and White 1978), or urban traffic management, show that policies can exist without their being the subject of major political debate. However, related issues – such as safety laws or transport investment – are likely to inspire conflict, since they strike more deeply at utilitarian interests.

When political salience does exist, however, the conditions may be set for ideological divisions, embodied in party political attitudes towards the want or ideal. At this point there is superficial attractiveness in Anthony Downs's (1957) assertion that 'parties formulate policies in order to win elections rather than win elections to form policies'. What Downs cynically ignores is the motivation of parties in the first place and their vulnerability when they formulate policies that *contradict* their ideologies, and also that most politicians hold and represent values and beliefs in addition to a simple wish to rule. In reality, when it comes to translation of wants and ideals, via the formulation of objectives, into policies or plans, it *does* matter which party is in power. That power extends to control over the agenda as well as to control over the resources for policy making.

The assertion that it is impossible to work as an applied geographer without taking account of ideals and wants is much less contentious now than two decades ago. Even if the objective of the researcher is apolitical – as it can be – the research itself, if it is to have any meaning, cannot escape either leading to judgements about past policies (or lack of policies) or the needs of future policies. Thus a simple statement of what exists (say, an estimate of housing stock) becomes a significant fact *only* when related to questions of what is desirable or undesirable, and to whom these evaluations apply. Without this addition the simple statement of fact has little meaning.

This is not a radical view in a political sense: it is not only marxists who recognise that spatial (as well as aspatial) distributions have a political meaning. It is, however, a realistic assertion, like that made by Hagerstrand, quoted on page 3, that applied geographers should know the political environment into which their work must take them. Whatever their own political philosophy they cannot escape the philosophy, ideals and ideologies of others.

Resources

The quantity, use, distribution and control of resources strike right to the heart of policy making. Indeed, it is difficult to conceive of any significant innovation that would not involve some change in these parameters. Resources and policy making share two relationships. Firstly, policies tend to be about resources; their division among different activities or groups of people; their depletion and conservation; their creation or acquisition. Secondly, the resources that are available inevitably influence policies. This applies at a variety of levels. In a regional plan it might be made clear from the outset that a certain sum is available for investment in public facilities. Thus a constraint is imposed on the plan. Or, in another case familiar in many less densely populated countries, the development of a region rich in (say) minerals, might be constrained by a shortage of other resources, for example manpower or investment funds. The Pilbara region of north-west Australia provides such a case (Hohnen 1976). Here the existence of very large iron ore deposits provides a major national asset which, at present, is mainly exported in the form of beneficated raw material. Economic analyses have indicated that the Australian economy would gain by developing a steel industry in the region. But to do this would require massive investment in social infrastructure and the attraction of a substantial population to a region which many white Australians find inimical.

An effective way of discussing resources in general is in terms of *stocks* and *flows* attributed to places (or regions) and to functions (or sectors). Such an approach has been proposed as an appropriate method for regional accounting by Czamanski (1972, 1973). Stocks comprise not only reproducible assets but also non-reproducible assets, financial capital or claims, natural resources and human resources. Flows consist of the transactions between different sets of stocks, or between stocks and consumers of stocks. In a strict accounting system, only those stocks and flows which could be given monetary or exchange values would be taken into consideration, but this is too inflexible for many of the interesting questions that arise when we are concerned with, say, environmental issues or complex long-term urban and regional plans.

The question of monetary values attributed to stocks and flows of resources is a complex one to which we shall return later, but which requires a short statement here. Ultimately in the planning or decision-making process there is a need for weighing or evaluating the advantages and disadvantages (or benefits and costs) of what is proposed. Monetary values *may* provide a convenient way of doing this. They permit us to generalise about opportunities that are being created or foregone and to make broad comparisons among alternative courses of action. However, it is not money *per se* that acts as a constraint or an objective of a plan (although very large financial sums can be impressive, if not actually frightening). Rather it is the *real* resources that are involved, and the wealth that is being created, or diverted from one use to another that is important. For

example, in a regional policy the real constraint is not so many units of money, but the availability of a stock of economic activities that can be relocated. In housing policy, the objective is not to create monetary value – which might as easily be done by restraining construction and causing inflation – but to create a larger stock of accommodation.

As part of the overall framework, resources are related not only to the central core of policy making, but also to the other three 'environmental elements'. Information systems require resources: even where censuses, for example, are well-established regular events they still require skilled management and analysis. In poorer countries, where trained manpower is a scarce resource, the problem of achieving an accurate census may be almost insuperable (Mabogunje 1976). Resource surveys are even more costly and demanding in human resources as well as equipment. Information systems are also resources in their own right, in so far as they facilitate the use of other resources. In this sense information has a value, like acquired skill. An important link also exists with institutions, particularly in relation to ownership and control and appropriation of resources. This applies not only to private ownership, but also to relationships between different levels of government and to different public agencies. Lastly, the essence of wants and ideals lies in the conflicts that arise over the distribution of resources.

Information systems

Ignorance and uncertainty are two of the most important limitations on rational and consistent policy making. Neither can be completely overcome but it is important to recognise where they occur and to work towards their reduction. This will not eliminate the debates and conflicts that surround policy making but at least it may put them onto a more informed plane, particularly if all sides have equal access to information. Moreover, we can assert that the more far-reaching and complex a policy (especially if it falls within the indicative rather than the regulative category – see p. 12), the greater is the need for well designed and full information systems.

Information systems, when properly constituted, have a number of uses. They help: (a) in *identifying* problem areas, whether they be environmental hazards or stresses, or imbalances and deficiencies in the use and distribution of resources; (b) in *forecasting* future conditions; (c) in *evaluating* the potential effects of proposed policies or decisions; and (d) in *monitoring* trends and the impact of current policies.

Not least, they may have a virtue of helping to point out areas of ignorance – a function of all good intelligence systems.

For applied geography the challenge of information systems comes in two ways. The first concerns the use of an understanding of the elements of urban or regional or environmental structures and processes (together with the needs of information users) to design an information system. The second is the transformation of knowledge, gained from empirical research and data collection, into a form that is usable in the structured framework of the system.

These are formidable tasks. Indeed, the establishment of workable *automated* urban, regional or environmental planning information systems has generally only been at a rudimentary level even among governmental agencies, where their relevance might seem greatest. Where systems have been developed they have tended to be limited to the simpler tasks of collection, classification and reproduction of basic data. For some purposes this is sufficient: for example, if one only wants a simple inventory of land use or occupancy this can be provided quite quickly by remote sensing methods and geo-coded areal units. More complex tasks, such as planning control or resource allocation, require information of different kinds to be brought together for analysis and interpretation and it is here that major problems emerge. At a higher level of complexity, such tasks as budgeting and resource allocation, or impact analysis and evaluation impose a need not only for information of a more intricate and sensitive kind than is normally collected, but also for systems with inbuilt 'feedbacks' between various sets of data.

Although information systems do not have to be automated they nevertheless impose a strict discipline in their compilation and use. This is true whether a system is 'general purpose' or 'problem-specific'.

The first type, the general-purpose system, demands consistency in its accounting units: grid squares, street blocks, buildings, and so on. Then, for every areal unit, information must be collected on the common set of variables and attributes. If the purpose is analytical or predictive the data should refer to the same point in time; if it is for planning control then the crucial feature is that it should be brought up to date continuously.

The second type of information system, problem-specific, might perhaps be better described as a system for assembling information for a unique purpose. That is, it does not exist as a continuing and updated set of facts, but rather as a structured set of questions to be used as occasion demands. There is a variety of examples, but one which has been the subject of much research in the United States is associated with the preparation of environmental impact statements, required by legislation whenever federal projects are being planned. Although different projects lead to different impacts, there has been considerable research to define systems for collecting, codifying and analysing information (Canter 1977). Less formal examples are found in other project evaluation methods that have been proposed such as the planning balance sheet (Lichfield *et al*. 1975) and the goals achievement matrix (Hill 1973). Neither of these specify the questions, but both propose strongly structured systems of arranging information prior to evaluation. Further discussion of these problem-based information systems is contained in Chapter 8. A third example is the preparation of 'programme budgets', which require well-structured monitoring and evaluation systems linked to the resource allocation process (Cutt 1978). This is discussed further in Chapter 4.

Thus what goes into an information system (particularly an automated one) is necessarily circumscribed by the needs for consistency and exhaustiveness. But it is affected also by other factors. One is the conceptualisation of problems – the definition of what is relevant – which is influenced both by theory and by social values. In a hypothetical example, an information system dealing with regional disparities in social welfare might emphasise the distribution of earnings or unemployment and pay less attention to negative externalities (or social costs) imposed on the general population by congestion or pollution in different areas, or to accessibility to economic opportunities. Another influence is the collectability of information. All information has costs and benefits and some is easier (cheaper) to collect than others. For example, traffic flows are easier to enumerate than reasons for movement, or choice of modes of transport, or social and economic effects of travel. We might therefore expect to find that the more readily collected information is, indeed, the most frequently collected and used, even though it is not necessarily the most appropriate. A third major influence on an information system comes from institutional factors which affect both the type of information available and the units for which it is available. This might be the result of rules on privacy and disclosure or it may be due to the information being derived from some pre-established base, such as census areas, or local government areas. However, it is common to find that basic information units are incompatible: census areas differ from water-supply areas, which differ from traffic zones, and so on. Unless one can go back to the smallest units – properties or households, for example – and 'reconstruct' areal units where necessary, there is no way of overcoming this problem satisfactorily.

Institutions

All societies possess rules, procedures and functions that govern the way they conduct affairs. Ideally such institutions would be formed by consensus and would facilitate actions by groups and individuals alike. However, some institutions circumscribe the activities of some people; and some eventually become obsolete encumbrances. Thus there are always likely to be pressures for institutional change: from the opponents of the 'system' and from its would-be improvers. Meanwhile, of course, the institutional framework must inevitably influence the manner of planning and policy making and, hence, the character of the landscape itself.

Two sets of institutional questions present challenges for applied geography. The first is the manner and extent to which existing institutions, and the division of powers and responsibilities affect the distribution and use of resources. A good example which has been documented by Linge (1967) and Logan (1979) is the effect of the Australian Constitution on the distribution of industry and the nature of urban and regional planning. The federal system limits the involvement of the Commonwealth government, and encourages a duplication of activities among the states. The second challenge lies in the means and extent to which geographers' research can contribute to the improvement of institutions.

Although national constitutions represent a scale that is not familiar to most geographers, there is a parallel to be drawn between their evolution and the manner in which planning and policy making have developed in recent decades. Constitutions often give the appearance of drab little documents full of fine phrases in legal jargon but having little relevance for living structures. In this the form belies the substance, for most constitutions are immensely powerful; their importance is reflected by the difficulties that have been faced by those trying to change them. An international classification of their content presented by Blondel (1969) explains why they possess such importance. In the earlier ones emphasis was upon the organisation of the State and on the separation of powers. More recent ones (post-1915) concentrated greater attention on participatory roles of groups in the population. Still more recently, constitutions devised since 1945 have tended to show a strong normative content with statements of the values and aims of the State. Blondel points to a historical transition from a simple declaration of rights to a stage where governments are expected to play a positive role in providing these rights and, hence, in promoting greater equality. In this shift, we see a growing interrelatedness of political institutions with political ideology and social theory.

At the urban and regional levels these issues are reflected in two main areas of concern. The more traditional one is most familiar in the work of political geographers on boundary problems arising from the vertical and horizontal separation of powers (Pounds 1972; Prescott 1965). Much of the earlier work in this field tended to ignore the conflicts of centralism and separation that underlie debates and disputes about boundaries and to adopt a mechanistic structural-functional approach (Kasperson and Minghi 1970). However, recent decades have seen an expansion of concern about the efficacy of boundaries and about whether they impede or facilitate planning and development. When these questions are raised, they are followed by questions about the economic, social and environmental values that underlie any particular arrangement of boundaries and the powers that they help to define. The different views on local government reorganisation reflect this: should local government have more or less power compared with central government? Should local government be in larger or smaller units? And so on.

The second area of concern stems from the earlier discussions of modes of planning and of the implementation of plans and policies, and is related to the nature of the powers and responsibilities held by different bodies. Answers to the questions raised here may have an important bearing on the character of physical and social environments. For example, a housing authority which has the power to allocate its tenants to particular homes may succeed, perhaps unwittingly, in creating a form of social segregation (Dennis 1978). Similarly, placing responsibility for a function with one government department rather than another can influence the way in which a resource is used: for example, allocating state forests to a Forestry department, rather than a Parks department, may lead to exploitation rather than conservation (Thier 1979).

Frameworks for private decision making

Although this chapter is primarily concerned with public policy there is also a strong case for devoting some attention to private decision making. Firstly, it is different (in kind, if not in principle) from the public process if for no other reason than it is dominated by personal, rather than social motives. The preceding discussion in this chapter is only partly relevant to private behaviour. Secondly, it is important in its own right. In mixed economies a large part (rarely less than 50%) of production and consumption is carried out in the private sector, and its actions are inevitably reflected in economic landscapes. Thirdly, much public policy is actively involved in trying to change private production and consumption characteristics. Hence it is important to know what it is that is under pressure to change.

One thing is quite clear: namely, that the normative assumptions about private behaviour found in neo-classical economies have little basis in fact. People are not omniscient utility-maximisers (Hollis and Nell 1975); nor are private corporations, although it has been argued (Hoover 1954) that organisations come closer to that abstract notion than do individuals. There is now sufficient evidence to show that corporations are far from omniscient (Green 1977) and that their motives are less concerned with maximising profits than with satisfying minimum acceptable conditions (Townroe 1971; Hamilton 1974).

It is also doubtful whether one can reduce private decision making to a single satisfactory framework. However, like the public policy framework above it is valuable to provide a broad classification of factors which need to be taken into consideration: at least this has a heuristic value. In this instance the environmental elements can remain the same, at least in outline, though their content might be somewhat different. However, rather than try to define a core sequence it is enough to identify broad elements which are generally involved in private decision making. These are perception, search and choice.

Perception
Although perception studies have their origins in psychology, their adoption and adaptation by geographers to include human mental responses to environmental and social phenomena has provided a rich area of research. Strictly, perception is a mental response to a physical stimulus, 'leading to the recognition or identification of something' (Drever 1967). But geographers have interpreted this definition liberally (Gould and White 1974); their laboratory has ranged from the neighbourhood to the continental scale and covered phenomena as disparate as natural hazards, urban pollution, residential quality and economic opportunity. Also, in addition to recognition and identification, they have considered 'preference' as part of the definition of perception. This is not without some justification, for perception is not just a matter of a reflex response to a flow of information. The response is inevitably influenced by a number of personal and social filters, such as variety of experiences and beliefs and values: often one sees (perceives) what one wants to see, or is conditioned by experience to see.

Relating behaviour to perception raises interesting practical questions. If one can define how and why people identify (and identify with) certain phenomena rather than others, or express certain preferences, then one can go some way to accommodating or influencing them. A basic question, derived from psychological experiments, is whether a respondent, faced with many bits of information (say, the entire distribution of shops and shopping centres in a city), perceives them holistically (i.e. in a *gestalt* fashion) or individualistically (i.e. in a structuralist fashion). Whichever is the case, there are distinct implications for individual behaviour, and also for policy making. In the environmental field, for example, there is a tendency to see actions simplistically, rather than as something to be accommodated within a complex system.

Then there is also the question of whether people perceive particular phenomena at all, or the degree to which they do so. Again environmental perception provides a rich source of evidence, with the same hazard being seen in a variety of ways by different respondents (Burton and Kates 1964). But there is also much to be learned from the degree of perception of policies. When it is important, for the fulfilment of a policy, that the private sector responds to certain kinds of information (be it coercive or inducive) then it is important also to ensure not only that the information is reaching its targets but also that it is perceived accurately.

Search
When a decision has been taken that it might be necessary to act upon some perceived want, there then follows a procedure of search. In trivial cases, or where decisions and actions are repetitive, this can be omitted. It might also be ignored in cases where

the object of possible search is known not to be variable, either in price or in quality. But this still leaves many significant actions in the lives of individuals, households and organisations which occur infrequently or irregularly and which do require a search process.

Conceptually a search process can be reduced to a cost–benefit equation. That is, search will continue as long as the *possibility* of finding a better item (be it furniture at a discount price, or a new house, or an industrial site) or a better method of doing something is believed to outweigh the *known* costs of further search. But this tells us little. The key questions are what gives rise to the belief that further search might yield greater benefits than those already assured, and what influences the costs of search. These can be answered in two ways. One is to identify factors related to the individual or organisation itself: the other is to examine external influences which act upon the decision makers.

In a gaming sense, some people tend at times to act as risk-takers and others as risk-avoiders. The reason may be psychological : for some people gambling can become a personal need, not limited to cards, dice or horses. But it can also be a rational part of a search strategy. If, for example, the search process can only work in one direction (i.e. there can be no going back to earlier finds), then whether one continues or not can be influenced by the resources at one's disposal and the time available before the search has to be discontinued. With these resources the risk of not finding a more satisfactory alternative will not be any greater or less, but it can be more easily absorbed. Conversely, if there is a premium on speed of search or only limited funds to conduct it, there is a greater propensity to stop at the first (or an early) satisfactory situation.

Costs also can be influenced by the nature of the searcher and the efficiency of his search. Unless outside sources are used (e.g. consultants) search tends to take place within the action space – or more familiar territory – of the searcher. This can be inferred from examples of actual behaviour: the tendency for migrants to go to places where they already have contacts (Cox 1972), and for firms to move within radial sectors of city regions (Sant 1975). It has also been used to explain the success of chains of retail and local companies: their identical format regardless of location may breed a sense of familiarity which saves the consumer from having to incur the costs of search and experiment (Relph 1976).

External influences on benefits and costs of search are easy to generalise but difficult to pinpoint. They stem from the efficacy of formal and informal information flows. Earlier, in the discussion of perception, it was suggested that information could be inaccurately perceived or distorted by the receiver. But the quality of the information sought or available to the decision maker may also vary. In many areas of search this may be overcome by the use of specialist consultants. Industrial location and marketing are two examples with geographical aspects. Using a consultant is a means of buying access to a wider information field, and probably to better information as well. But consultancy services are rarely free and their use is determined by the same cost–benefit framework as any other aspect of the search process. On the other hand, certain kinds of information service are free, at least at the time of consumption. These include formal services established by the public sector to increase the impact of policies. Some are overtly spatial or regional, such as the Location of Offices Bureau in Britain. Others are more concerned with productivity, like agricultural extension services, but indirectly have an impact on the economic landscape as well.

Choice

Normative neo-classical economic models treat the exercise of choice as a rational expression of personal utility. The benefits of having a good, or so many items of a good, are traded against the benefits of having another good and welfare is maximised when a person is indifferent to any arrangement of expenditure other than the one he has chosen to make.

This approach to the explanation and prediction of choice has much to commend it, as long as one realises its limitations. Firstly, variations in perception and search behaviour belie the ability to make such fine-turned decisions. Secondly, it is not always clear, without intensive research, what is being traded. Thirdly, there may be limitations on the things than an individual actually can trade.

Identifying what it is that is being traded is important both for the development of hypotheses and for the practical issues involved in evaluating the impact of plans and projects. Often a trade-off is relatively straightforward, involving a choice between one item and another. However, among major decisions (such as are embodied in land use) the choice may be between 'bundles' of items, with each bundle containing interdependencies that are difficult to disentangle. Residential choice provides an important example where it is necessary to guard against oversimplification. The familiar model, where choice is based on a trade-off between expenditure on land, the price of which is seen to be related tó distance from a city centre, and expenditure on travel (e.g. Alonso 1964), is not wholly realistic. People may spend the same amount on accommodation (house and land) regardless of whether they live near the centre or in an outer suburb, and the real trade-off may be in lifestyles rather than in transport costs. If this is the case the implications for the evaluation of, say, the impact of a new transport scheme takes on considerable complexity, because costs and benefits no longer can be attributed to single goods.

In addition, there are limitations on what people *can* trade off, which raise important questions for geographers. In welfare economics an evaluation of the utility which an individual derives from a good is defined by his *willingness to pay*. If the consumer is willing to pay more than he actually has to, then he not only derives the benefit of having the good, but also gains an extra benefit referred to as 'consumer surplus'. Clearly, if we had a free choice in patterns of consumption we should aim to maximise our personal surpluses. But such freedom is not equally available to everyone. In reality *willingness to pay* must be modified by *ability to pay* and *opportunity to buy*.

The first of these terms, which refers to inequality in real disposable income and wealth, is familiar in critiques of welfare economies but it also has geographical implications. Variations in ability to pay affect the spatial distribution of choices. This relationship was stressed by Pahl (1971) in his social analysis of rural settlements, in which he asserted the existence of a dichotomy based on wealth. The affluent placed a high value on the space and amenity offered by low-density rural living while the poor saw the same environment as a constraining influence. In effect, the latter were trapped by their poverty – unable to afford urban house prices and yet unable to accumulate wealth where they were living, and suffering at the same time from poor access to services. Economic environments abound with real and potential 'entrapments' which deserve imaginative research. People living in hazardous or unattractive locations may do so not out of choice, but because they have no effective means of escaping; others who use their agricultural land less productively than the optimum might do so because they lack the funds for investment or the means of reorganising their systems.

Opportunity to buy limits freedom of choice in two ways. The first is in the degree to which goods are appropriatable for personal use and consumption. Some goods are 'free' in the fundamental sense that they cannot be bought and sold but are there for the enjoyment of all – though this may need to be reinforced by law to stop despoliation. Others may either be made free or allowed to be appropriated, depending on the law: for example, a foreshore may be held as public land or it may be allowed to become privately owned. At a more familiar level, opportunity to buy (and lack of it) is embodied in land-use plans which regulate the location of activities by designating uses that are permissible in an area. The second aspect of opportunity to buy is related to the production process. Unless one is able to afford a 'custom-built' commodity or to produce it oneself, one has to take what is offered, and this might be tailored to an image or stereotype that fits only a part of the market. Examples are provided by the suburban housing built around many western cities, which often conforms to the notion of the standard family and to what the housing finance market finds more acceptable.

Summary

Public policy making and planning take place at a number of *scales* from the local to the national. They concern subjects as disparate as physical structures and social conditions and relationships, and have a wide spectrum of complexity. They can also take place in political environments which favour different modes and intensities of public involvement.

Given these various dimensions it is necessary – if one is to be more than self-indulgent – to have a view of the broad context within which policy and planning take place: what the processes of decision making are, how they are affected by 'external' conditions, how public and private systems can be reconciled, and so on. There are many different ways of looking at these. None are perfect but, to be effective, it is necessary to have a framework which takes adequate account of the limitations and opportunities that surround policy makers – or what Hagerstrand (see p. 3) called the 'political dimension of human affairs'. While striving for objectivity in ourselves we also need to understand that actions are spurred by subjective motives, especially among those with the power to effect change.

3 Wants and ideals

An allegory

In locational studies there is a familiar example (which we can modify slightly) that is used to illustrate the nature of spatial equilibrium under hypothetical conditions of monopolistic competition. The scene is a beach, evenly covered by holidaymakers, each with the same perfectly inelastic demand for ice-cream which is sold at a fixed price. Enter two profit-maximising ice-cream vendors. Where will they locate? In this case, originally discussed by Hotelling (1929), the answer is determined by the nature of the market. The ultimate equilibrium location for both vendors is side-by-side in the centre of the beach. Anywhere else is certain to lose custom for one, unless after collusion they decide to locate symmetrically along the beach. However, in this example there is no benefit to either vendor from collusion since the total volume of sales would not be affected and their shares would be the same as at their central location. On the other hand, there would be a potential benefit to consumers, since they would have to travel shorter distances to the sellers. Their gain does not come from a change in consumption (the assumption of perfectly inelastic demand prohibits that) but from the reduction of travel time which can now be used for other purposes. Moreover, since ice-cream is a perishable good, their greater accessibility means that the product will be in a better condition after it has been brought back from the kiosk to the spot on the beach where it is to be consumed.

Thus, in the absence of collusion and with the market operating freely, there is a wide disparity in access to the point of supply. With collusion the market is no less (but no more) profitable to the suppliers, but consumers reap some benefits from greater accessibility, though there are still disparities in access to suppliers. Short of having a kiosk beside every customer there will always be some such disparity.

Now let us introduce another group to the beach: non-consumers of ice-cream, who arrive before the vendors and also distribute themselves evenly along it. Eventually the vendors arrive at their central equilibrium location. In pursuit of sales they play loud advertising jingles. And in pursuit of profit they deliberately cut down costs by not providing garbage cans. The result is another set of disadvantaged people in addition to would-be consumers located furthest from the kiosk. The non-consumers in and near the centre have to absorb a reduction in welfare whether they stay and endure the inconvenience or whether they incur the 'cost' of moving to more remote parts of the beach.

This trivial example illustrates two aspects of welfare arising from the location of consumer services: inequalities in accessibility and the imposition of negative externalities. But the moral of the story is that the extent to which these occur is neither fixed nor inevitable but depends, instead, on the 'rules of the game'. It is possible to alter the rules for the case of the ice-cream vendors in a number of fundamental ways. For example:

The number of vendors could be changed.
The locations of vendors could be fixed (by law).
Vendors could be constrained by environmental legislation to provide for garbage disposal and to refrain from noise.
The beach could (in theory, at least) be divided into zones for consumers and non-consumers of ice-cream.
The more remote consumers could be subsidised (e.g. given free ice-cream) to make up for their lower accessibility.
The ice-cream industry could be nationalised, and made to operate as a non-profit-making social service, with ice-cream freely available on request to everyone.

These do not exhaust all the possibilities but they are sufficient to suggest ways in which distribution of costs and benefits arising from the location of people and activities can be altered merely by changing the rules which govern the market. But in changing the rules some people who were better off before are likely to be made worse off, and vice versa. Inevitably, if this is the case, there will be conflict: people will argue for what they want for themselves or for what they believe to be 'right' for others. There might also be conflict if some people, without being made worse off, see others becoming better off. As Runciman (1972) has argued, conflict may as easily be prompted by relative deprivation as by absolute deprivation. If a change does eventually take place then, clearly, one set of wants (and, possibly, ideals) will have been substituted for another set.

Thus, when we talk of 'changing the rules of the game' we immediately have to deal with two major questions: How are wants and ideals formed and expressed? and how do people respond when their wants and ideals are *not* met?

Answers to the latter question help to shed light on the former. Basically there are three main options open to a person who is dissatisfied with his living or working conditions and opportunities and unable to improve them by personal action. Not all of them may be equally available at any one time, and in many circumstances it would be unrealistic to consider them as equal alternatives. The three options are: to stay and endure the conditions; to escape the conditions by moving; and to attempt to change the conditions. Strictly, only the third is directly concerned with 'changing the rules', although the second may evoke a response by those who stay and find *their* conditions being worsened. This is in keeping with the observation that declining regions have their problems exacerbated by selective outmigration of people seeking better economic opportunities (Richardson 1978). It is in the interest of those who remain to promote regional development policies with assistance coming from elsewhere.

'Attempting to change conditions' in this context means engaging in political actions whose purpose is to bring pressure to bear on the distribution and use of resources. The means of doing so are very varied and differ from one situation to another. Some appear in the shape of formal institutions; others are informal, based on loose alliances of interests; some work at the local level; others on a national scale. Identifying and designing such institutions are practical issues that have provided a rewarding field of research among geographers (Massam 1975). However, it is not so much the structure and function of political action that are of interest here, but the generalisations that can be made about the form in which pressure groups express their wants and ideals and the geographic implications that stem from this. We do this by turning next to a discussion of private, public and merit goods.

Private, public and merit goods

The distinction between 'wants' and 'ideals' is that the former refer to *immediate* needs or desires capable of being supplied by an appropriate disposition of resources, while the latter refer to *ultimate* objects of endeavour conforming to some standard of excellence or state of perfection. Wants are expressions of self-interest (though some may only be attained by group action); ideals are expressions of community interest which transcend the needs or demands of any single individual. Clearly the two are fundamentally different even though one (wants) may emanate from the other (ideals). However, this twofold classification is insufficient. It is, firstly, necessary to make an additional subdivision between two types of wants: *private* (or personal), and *public* wants. Secondly, we should seek other dimensions on which the differences between wants and ideals can be elucidated; these are summarised in Table 3.1, which is partly based on Allison (1975).

Private and public wants and ideals can be juxtaposed against modes of production and distribution which we can refer to, respectively, as private (or market) goods, public goods and merit goods. The first of these, *private goods*, is associated with

Table 3.1 Summary of the characteristics of private and public wants and ideals

Social context	Private wants	Public wants	Ideals
Modes of distribution	Market goods	Public goods	Merit goods
Decision-making powers	Exchange	Bureaucracy/Polity	Polity
Units of exchange/ value	Cash	Cash/Votes	Esteem
Ideological base	– Utilitarianism –		Utopianism
Motivation and political support	– Interest groups –		Principle groups

market mechanisms, cash transactions (and, hence, with value systems derived from the existing distribution of income and wealth) and private ownership. Its base lies firmly in the utilitarianism of classical economic theories of value and distribution in which individual preferences are satisfied by consumers purchasing goods and services up to the point at which they are indifferent to any further consumption. Left to itself, this (hypothetical) system works towards a Pareto optimum in which social welfare is maximised through the process of *voluntary* exchange. Note the word 'voluntary'; in the neutral world of economics it makes no difference what is produced and consumed. As long as it is done willingly and reflects preferences or utilities, it contributes to welfare.

Very few people still hold the view that the market is a perfect medium for allocating resources or distributing welfare (though many still regard it as the best available medium). However, without departing from the utilitarian base it is possible to conceive of other arrangements (i.e. *public goods*), whereby either production or consumption, or both, are carried out collectively. These tend to be items which can either be produced or consumed more efficiently if done collectively, or which contain potential discrepancies between private and social costs and benefits. The latter means that a person gains or loses gratuitously as a result of the expenditure undertaken by another. Transportation provides examples of both private goods (cars) and public goods (urban roads). (It also contains merit goods, in the form of subsidised transport, provided where the market is insufficient but where accessibility would otherwise be impaired.) Decisions still have to be made about how much of each good to produce but these, instead of being determined by the summed utilities of individual consumers expressed in their patterns of expenditure, are now determined by the estimation of social costs and benefits by a single decision-making body, called a 'bureaucracy' in Table 3.1. Money – as a measure of real resources and opportunity costs – is still the main measure of value of a public good, as it is for a private good. However, since a public good is ultimately controlled by the polity, through elected governments, it is reasonable to suggest that 'votes' also act a unit of exchange. That is, if the community is unhappy about the amount of resources (too much or too little) given to the provision of a particular public good, it also has recourse to political action.

Turning to the third category, ideals, it is apparent that the utilitarian base of the other two categories no longer holds. Ideals have their origins in intellectual constructs of man's 'proper' relations with his universe and with himself. To use the terminology of utilitarianism, welfare maximisation is not the summation of individualistic or collective utilities, but the promotion of particular patterns of behaviour and consumption that are believed to be superior to others. Needless to say, ideals come into conflict with other ideals, as well as with the various kinds of wants. Just as each person has his own set of 'utility functions' (or ranking of wants), so each also holds a different set of ideals. A further difference is that ideals are based upon *principles* and wants are founded on *interests*. Thus, following Allison, we can talk of principle-groups and interest-groups to describe the parties in a conflict.

At this point the concept of a *merit good* must be reintroduced, for it is this that gives practical and normative expression to an ideal. That is, a merit good refers to the treatment of a particular good whereas an ideal refers to much broader conditions than be encompassed in a single good. For example, such concepts as 'full employment', 'optimum settlement structure', and 'zero population growth' all represent ideals, but at levels of abstraction that need to be reduced to manageable proportions. This can be done through the concept of a merit good, which is a means of pursuing an ideal.

Merit goods are defined as those goods which, compared to some criterion, *other than utilitarian ones*, would be under-consumed if the market was left to operate freely (Musgrave and Musgrave 1976). The definition of under-consumption requires some discussion, for it is not one that can be derived logically. In effect it means that, on the basis of the specified criterion, the market equilibrium based on prevailing conditions of supply and demand leaves some people not consuming any of the good or consuming too little of it. The most familiar example is education where, clearly, the motive for the level of provision actually made in many systems is a better educated, more civilised society rather than some cost–benefit calculus: otherwise why subsidise students to read such esoteric subjects as ancient languages, moral philosophy (or even geography)? This, and many other examples, possesses the attribute that market forces are likely to result in not enough of the good being produced or consumed.

Parenthetically we might also think of merit 'bads' (following Smith 1977) which are those things that are over-consumed and, hence, detract from the ideal: largely these are the obverse of merit goods: pollution, destruction of natural species, congestion, slums, and so on.

A further point that should be made before proceeding is that we have deliberately avoided the concept of *need*. The reason for doing so is that it is subsumed by the concepts of wants and ideals and that these offer more precision and better links to other concepts. Needs, as discussed at length by Smith (op. cit., pp. 27–39), cover the gamut from basic human physiological processes of survival, through personal psychological needs (fulfilment, pleasure, happiness) to social and organisational needs embodied in welfare and defence. Moreover, while 'need' may in fact underlie the concepts of wants and ideals it can only be activated by using the terminology of these concepts.

Although the definition of a merit good places it above utilitarian values, it is impossible to escape completely from utilitarian concepts. Some explanation has to be made of *why* certain goods are over- or under-consumed. Also, in specifying the means whereby a merit good can be promoted and the amount by which it should be increased (or decreased, in the case of a 'bad'), one inevitably confronts resource allocation questions which can also be treated from a utilitarian standpoint. Indeed, this is a continuing source of conflict between various interest groups and principle groups. At its broadest, the conflict concerns the importance of traditional economic goals. The issues are not about production *per se*, but about what to produce and how to distribute the rewards according to criteria other than utilitarian ones. But in order to conduct the argument one must expose the effects of utilitarianism, and ask why it is inadequate. It may even occur that this process will lead to an ideal being modified if it cannot be sustained against utilitarian arguments. There are no *a priori* grounds for accepting that an ideal promoted by a principle group will, in fact, lead to the welfare of the community being improved.

Under-consumption may arise from *conditions of supply*. It has long been understood that monopoly and oligopoly result in long-run equilibrium levels of output that are smaller than that achieved under perfect competition and that this also extends to conditions of spatial monopoly, especially if reinforced by f.o.b. pricing (Greenhut 1956). Spatial 'protection' is particularly important where scale economies and other forces for agglomeration are associated with a tendency for goods and services to be supplied from a decreasing number of locations. Where this is accompanied by consumers having to absorb the cost of travel or transport, there is likely to be a disparity in consumption based on distance from the point of supply. This applies as much to public goods and services as it does to private ones. Such effects have been documented for a number of cases, such as health services (Haynes and Bentham 1979) and recreation facilities (Patmore 1973). Other examples may occur where there is not the same element of spatial monopoly and protection: here under-consumption becomes a more general phenomenon based on income. For instance, the operation of a monopolistic or oligopolistic transport system (such as Australia's two-airline policy for internal travel – Hocking, 1971) can result in higher fares, fewer routes and hence lower accessibility between places and people, and less travel. Some might argue that this helps to prevent over-consumption of a scarce resource and creates less disturbance from aircraft. Others might say that commercial, social and familial relationships suffer from the 'tyranny of distance' (Blainey 1966).

'Conditions of supply' means more than production functions, intensity of competition, and pricing policies of firms. In many societies ownership and control are increasingly divorced as growing shares of national economies are absorbed by large corporate organisations. Both private and public sectors alike have been the subject of 'bureaucratisation' (Marris 1974). When this happens the possibility exists that a bureaucracy will develop a set of motives and modes of behaviour of its own, and that what Mesthene (1974) calls an 'ideal–real gap' will evolve, in which the bureaucracy ceases to be responsive (or becomes less so) to wants or ideals expressed in the community. Gillingwater (1975) has directed this criticism at urban and regional planners, suggesting that they tend to stress comprehensiveness and functionalism at the expense of flexibility and responsiveness in the management and planning of human and physical environments.

On the *demand* side, commonly cited causes of over- and under-consumption are low incomes, ignorance and social fragmentation.

Incomes have a crucial role to play since they are major determinants of propensities to consume and also of modes of consumption. Low-income households are not only likely to consume less than richer ones but also they may consume less 'efficiently' in certain respects. For example, in an inflationary housing market it is usually more efficient (for the consumer) to buy rather than to rent. The former action leaves him with an appreciating asset, often for the same outlay. But entry to house ownership is not equally easy for everyone.

Ignorance has a significant effect in many areas where short- and long-term costs and benefits are unequally known or evaluated, and where risks are underestimated. Examples are common in such areas as environmental health and safety and response to hazards (White 1974; Pearce 1976). Social fragmentation, which is taken very broadly to mean both intended and unintended lack of beneficial interaction between individuals, groups, or communities, works in two ways. One is to reduce the opportunities for collective consumption which would provide economies of scale, or even the necessary threshold to support an activity: e.g. home drinking has seen the tragic demise of the country pub in many areas. The other effect is to reduce opportunities for reciprocal investments whereby individuals derive mutual benefits from each other's expenditures (or vice versa in the case of 'bads'). As a trivial example, the inhabitants of a 'best-kept village' competition usually find that they all benefit from increased property values. In contrast, a street full of obsolete tenanted homes owned by absentee landlords is unlikely to win anything.

It would be tempting, but misleading, to present the three categories (private, public and merit goods) as a spectrum of modes of production and consumption fully consistent with political ideologies towards the distribution of resources and wealth. Without denying the existence of ideologies, we find any such generalisation difficult to sustain throughout our classifications. It is true that there are broad and profound differences between proponents of private market systems and whose who favour the socialisation of production, but this is a cleavage which, especially in mixed economic systems, is difficult to relate to the conflicts that arise over particular issues. Every issue, by definition, has at least two protagonists. They may both be private interests: for example, a private firm opposed by property owners over the issue of pollution. Or they may both be idealistic principle groups: for example conservationists versus a charitable housing trust in conflict over inner-city renewal and accommodation for low-income families. Or they may be any combination of the three categories. We may also find private interests arguing, quite rationally, for an extension of public goods (e.g. more investment in infrastructure) where this is seen as contributing to greater private benefits (or profits).

However, there is no great disadvantage in not being able to make a clear association between political ideology and positions taken on any single good. As long as we bear in mind the three-fold typology of private and public wants and ideals we are in a position to make better analyses of the distribution and allocation of resources and better evaluations of the effects of different actions.

Setting objectives and evaluating alternatives

The link between ideals and wants and the creation of plans or policies lies in the establishment of objectives. These are the targets, or thresholds, which the plan-maker hopes to achieve and without which there would be no reason for action. But translating wants or ideals into a formal statement of objectives is not necessarily a simple task. This depends on what is being planned, and by and for whom it is being planned. Simplicity of task and consensus about the need for it allow a relatively straightforward list of objectives. In contrast, a complex task, with a number of effects following upon a heterogeneous population, gives rise to some difficult problems in setting objectives. The opposite ends of the spectrum can be characterised, at one extreme, by single-sector private firms, pursuing a few *internalised* objectives, and at the other, by multi-sector public agencies pursuing many varied *external* objectives. In between there is a wide range of organisations with differing degrees of internalisation and complexity of objectives. The trend, however, appears to be that external considerations are increasingly imposed on organisations. Private organisations, in certain circumstances, have to meet external standards; public ones which previously had greater autonomy, have new operating objectives which extend their accountability. For example, in recent years governments have added to the responsibilities of departments concerned with primary production or utilities (e.g. forestry and water supply), requiring them to provide for multiple use (e.g. allowing recreation) and to observe conservationist principles. This may create for them a problem common in multi-sector agencies pursuing varied external objectives. They may try to accommodate as many of a community's wants and ideals as possible, but they cannot be all things to all men. The best of plans is likely to neglect some interests or principles, even if it does not actually contravene them.

Objectives are formal statements, even if (as is illustrated below) they are sometimes worded in fairly vague terms. This vagueness is, in fact, usually deliberate: it may be a way of implicitly allowing private interests to have some scope for manoeuvre, or it may be a device for retaining flexibility for the planning authority. An example, from the West Midland Regional Study Group (1971), is as follows: 'To provide within a satisfying total, a range of housing in accordance with future assumed population characteristics.' Such a statement is open to widely different interpretations. However, while it is not a firm target it is, none the less, a statement against which alternative plans could be evaluated. One land-use plan might provide for a wider range of residential environments than another.

In examining the objectives for a public multi-sector plan or project, therefore, we are likely to find a composite statement, made by a single agency, perhaps reflecting many disparate (possibly conflicting) interests and principles, in which some objectives are expressed in precise numerical terms and others are worded vaguely. This may seem like a prescription for chaos. However, a close look at the list of objectives preceding planning statements usually reveals categories of objectives that are consistent with earlier discussions of wants and ideals. Broadly the three categories are embodied in sets of objectives which stress: (a) economic productivity and private consumption; (b) efficiency in public sector provision; and (c) equity and protection of environmental amenity.

The first and third of these are illustrated by two objectives contained in the Sub-regional Study of the Coventry City Council (1971). At first they appear to present a direct conflict. However, the possibility of spatial separation reduces that risk, without fully obviating it. The examples are: (a) 'To locate new residential development in areas of high environmental potential'; and (b) 'To locate new development so as to conserve areas of high landscape value'. If we ask what is served by the first we must conclude that, on balance, the main parties to gain would be private interests – those to whom higher property values would accrue from such locations. The rest of the population stands to gain nothing from such development – and may actually have their access to, or enjoyment of, such areas reduced. On the other hand, provided access to areas of high landscape value is not restricted, their conservation (the second

objective) potentially confers benefits upon the whole population. Since private market forces would probably see these areas consumed for residential development (hence giving under-consumption of the experience of fine landscapes) this objective reflects an idealistic promotion of a merit good.

Unfortunately, consideration of every possible objective is an impracticable task, unless the organisation is a simple structure and can internalise its objectives. In public planning this is unlikely to be the case, and it is necessary to provide a filter between the community and the plan. That is, a person or a group of people have the responsibility for drawing up the list of objectives. This can result in bureaucratic unresponsiveness to community preferences, but planning legislation in a number of countries has attempted (not necessarily successfully) to safeguard against this by providing for formal channels of public participation.

In practice, planning teams have tapped a number of sources in order to derive their statements of objectives. These include surveys of community attitudes, both directly and through elected representatives or the spokesmen of pressure groups. They also include the opinions of planners themselves, who can combine their own experiences with their rationalisation of the needs of an area. This might take the form of a forecast of future trends which, when juxtaposed against established planning standards and statutory requirements, provide the basis of an objective. An example is given in Coventry Sub-regional Study referred to above. Having forecast a population of 1.41 million in the region, the study set as an objective the provision of 'sufficient land . . . at appropriate space standards and with all supporting services'. In this case, 'space standards' refers to residential population densities, plus land for industry, infrastructure, recreation and so on, all of which can be estimated by rough 'rule-of-thumb' methods according to conventional planning practice. 'Supporting services' includes items that are statutorily required, such as water supply and sewerage. In this instance we have what was termed an *essential* objective which could be derived without recourse to public opinion.

The term 'essential' objective is an interesting one, for it implies that an absolute need exists. In one sense this is true: everyone needs shelter and other means of subsistence. However, the *quality* of shelter and the *level* of living are not absolutes, and this is recognised in the wording, which refers to *appropriate* space standards. Nevertheless, a distinction can be drawn between an essential objective – which has to be fulfilled in some form or other – and *discriminatory* objectives, which may or may not be fulfilled but depend, rather, upon the extent to which they contribute towards the best overall plan. This point is illustrated in Chapter 8 when evaluation techniques are discussed.

Another difference between essential and discriminatory objectives in public planning is that the latter are less satisfactorily drawn from the experiences and rationalisations of planners. At this point it is necessary for them to identify the wants and ideals that exist within a community. Possible techniques for doing so are discussed in the next section.

The corollary of using wants and ideals to define the objectives for a plan or a project is to use them also in the evaluation of alternatives. What one seeks in such an evaluation is the 'best' line of action: that is, the one which comes closest to meeting the aggregated set of wants and ideals expressed in the initial objectives. There is a simple logic to this: why have the objectives if they do not guide the choice of action?

However, the issue goes further than this. A set of wants expressed by one person represents that person's optimum mix of goods. The same would apply to a community if we could similarly identify a single set of wants. Thus progress towards meeting those wants is the same as achieving a higher level of welfare. But if, as is almost certain, the community does not have a homogeneous set of wants then the task becomes one of not only judging progress towards each set but of comparing one set with another. In an economic evaluation this comparison is conducted through the use of monetary values. But, as we describe in Chapter 8, this is not the only possible way of evaluating alternatives.

Attitudes and behaviour

We know intuitively that within a community there are as many preference functions as there are individuals. It would be impracticable to attempt to identify all of them. Even if we were not daunted by the task we should certainly be frustrated by people's inability to make up their minds. It would also be frustrating to find that we could not necessarily identify all the people who would be affected by a plan: unborn children, and potential migrants to a region would not be represented.

Yet if planning is to be responsive it is necessary to identify the wants and ideals of a community and, notwithstanding the problems that might arise from parsimony, to do so in an economical manner. In effect this means using an *inferential* approach to deduce what the members of a community seek in a plan and what values they put upon their wants and ideals. Within this approach there is a variety of techniques. Broadly, these can be divided into two major groups, namely: (a) *behavioural* studies that attempt to infer attitudes from behaviour; and (b) *attitudinal* studies that attempt to predict behaviour from attitudes.

Leading examples of the first are found in normative studies of location, land use and interaction. These are founded upon the assumptions (and observations) that: (a) people aim to maximise the satisfaction of their private wants; and (b) these wants can be expressed by downward-sloping demand curves which reflect the marginal utility of consuming extra units of a good at different prices. In geographical analysis these prices incorporate the costs imposed by distance. An example is illustrated in Fig. 3.1 in which the gravity model is used to derive a demand curve (or 'willingness-to-pay' curve) to use a free facility – namely the English Lake District (Mansfield 1971). Since entry to the region is free, the only factor taken into account is travel cost. As can be seen, this is related to distance and reflected by the falling proportion of trips per 1,000 population as one moves further from the Lake District (Fig. 3.1(a)). In fact the relationship is not perfect because: (a) distance is not the only influence on cost (one might also include travel time); and (b) some places have access to other parks and recreation areas which divert demand from the Lake District. Mansfield's analysis continues by assuming that the cost of travel to use the Lake District at that point where the number of trips falls to zero defines the amount that all users would be willing to pay. From this it is possible to derive a value of consumer surplus: the amount that users would have been willing to pay but did not actually have to pay because their trip was shorter and less costly. From this one can derive the total value of consumer surplus (labelled CS in Fig. 3.1(b)).

In this instance we have the definition of a want based on inferred willingness to pay and we also have an estimate of the benefits derived from the fulfilment of that want based on the notion of consumer surplus. If we assume, temporarily, that the Lake District has an unlimited capacity for tourists then we

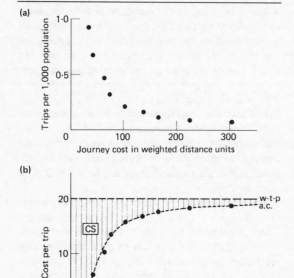

Fig. 3.1 Estimating the value of recreation facilities (based on Mansfield 1971)
w-t-p: willingness to pay
ac: actual cost
cs: consumer surplus

can go further. With a little more information on the determinants of recreation travel we could make an estimate of 'unrequited wants': that is, the number of trips that people would want to make, but do not because of the cost of travel or because of their low income. This could then be used in setting the objectives of a regional plan for tourist development. At the same time the information would provide the basis for evaluating the plan, by calculating the new value of consumer surplus and subtracting the costs of development.

Some of the problems in this approach (e.g. the measurement of cost) are discussed in Chapter 8, but there are two which impinge directly on the present issue. Firstly, there is the question whether the costs identified (assuming them to be accurate) really do measure willingness to pay. Remember that it is the cost at which the number of trips falls to zero that is critical in the analysis. Some people might have been prepared to pay more than they actually did without reducing the number of their trips. This cannot be inferred from an aggregative analysis, and hence it is

difficult to identify the strength of their wants. Secondly, it is also difficult to say whether wants are stable. If, on the basis of the inferred pattern of wants, a change occurred which made the recreation area cheaper to reach and thus encouraged more tourists, some former tourists might be so affected as to reduce their trips. Unless this can be predicted, and unless the pattern of encouraged and discouraged tripmakers is also known, an accurate evaluation may be impossible.

A second set of examples of attempts to infer attitudes from behaviour occurs in studies which try to explain 'gaps' between ideal (or optimal) behaviour and overt behaviour or revealed preferences. If such a gap does exist, it may be attributable to a number of factors. It may be due to ignorance or eccentricity on the part of individuals. It may be that the ideal, established by the analyst, is unrealistic. But it may also be attributable to a deliberate trade-off by an individual between the various benefits which might be derived from different patterns of behaviour. Thus, for example, one might postulate an optimum pattern of shopping behaviour in which people use the closest, cheapest store for each individual commodity, only to find that in reality people do not conform (Amadeo and Golledge 1975). If we follow up this 'deviance' we might find other things that shoppers value, besides proximity and cheapness. Similar gaps occur in most areas of human behaviour: examples include agricultural productivity (Wolpert 1964), and residential choice (Bourne 1968).

Unfortunately, it is extremely difficult to explain why a gap occurs unless one penetrates beyond the pattern of overt behaviour. It is not sufficient to rely upon an analysis of the character of the individuals concerned: one also needs to know something of their attitudes, perceptions and preferences. This can be illustrated with a simple gambling model from Hamblin (1975). In this there are three objective elements (see Fig. 3.2): the 'ante' that the gambler puts up (A) (which corresponds to overt behaviour); the amount that he stands to win (W); and his probability of winning (P_o). In an optimising

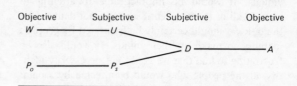

Fig. 3.2 Elements in decision making

situation A will be perfectly correlated with the multiple of W and P_o. But between these two sides of an equation we also have two sets of subjective elements which influence the relationship between the objective elements: the disutility of the sum that may be lost (D), the utility of possible winnings (U) and the subjective probability of winning (P_s).

When these intervene, the correlation becomes less than perfect. Moreover, their intervention has unequal effects on W and P_o. In Hamblin's experiment A was shown to be more influenced by W than by P_o (which we could have learned from any successful bookmaker).

The implication of this simple illustration is not only that it is difficult to infer attitudes (wants and ideals) from behaviour alone, but also that the opposite is equally true (Fishbein and Ajzen 1975). In this case we would be using U, P_s and D to predict A, and while the first three elements may describe a person's attitudes or perceptions they say nothing about his *intentions*. This is a shortcoming that reappears in much of the geographical research on attitudes. Yet despite this, it is valuable to elucidate the phenomena and the environments that people value.

For simplicity the second major group of inferential techniques, attitudinal studies, are presented in three broad categories, as follows: expressed preferences; willingness to pay; and conflict analysis. The second of these is an extension of the earlier discussion on the behavioural inference of willingness to pay.

Expressed preferences, related to attributes in physical and human environments, have been the subject of a large number of surveys using a variety of techniques. They have in common the procedure whereby a respondent to a survey ranks or scores places or phenomena according to his own preferences. Early examples of this approach were carried out by Gould and have been replicated and refined in many other studies (Gould and White 1974). In the earlier studies the objective was to elucidate residential preferences on a regional scale. Respondents were asked to rank areas according to their view of how attractive they were. The results tended to show strong preferences for the 'home' region, regardless of where the sample of respondents lived, plus a consensus about the attractiveness of other regions. From the first, one can infer that familiarity with an area, together with social linkages, lead to local loyalty. The second is more difficult to interpret: preferred areas tended to be environmentally attractive or to have economic strengths (high wages, low unemployment, variety of jobs) which could lead us to infer that apart from familiarity people also want security and pleasure. Later studies have probed a little more deeply, by asking not only about residential preferences but also for assessments of regions according to other attributes such as physical environment, economic opportunity and political attractiveness (e.g. Weinand and Ward 1979).

Methodologically, the most important extension to this approach lies in: (a) elucidating the motives underlying the expressed preferences; and (b) relating preferences to intentions (and, thence, intentions to behaviour). The first of these has attracted more attention than the second, and attitudinal studies have increasingly been accompanied by questions related to the characteristics, information bases and experiences of respondents. At the same time there have been numerous experiments with scaling techniques, with the intention of being able to penetrate more effectively into the description and explanation of attitudes. The use of such psychometric scaling techniques as Guttman's scalogram, and Osgood's semantic differential (described by Fishbein and Ajzen 1975, pp. 64–78) are becoming increasingly common.

A simple variation of deriving expressed preferences which has been used in planning studies is a checklist technique. In this respondents are 'given' a fixed amount of resources to be spent on a list of items which is shown to them: they may add items to the list if they wish, but need to be encouraged to do so. How the respondent allocates the total sum is then interpreted to be an expression of his preferences within the context of that plan.

Surveys of willingness to pay (the second set of techniques in attitudinal studies) are similar to their inferential counterpart discussed above, in that they relate to attitudes towards individual goods. Where they differ, however, is in their concern for precision in defining willingness to pay and identifying its correlates. The latter is particularly important where the distributional effects of a plan or project are to be taken into account. Surveys also take on added significance where other sources of data are unavailable.

Willingness to pay is an elusive concept, for it is related to ability to pay and need to pay as well as to individual preference. It is not uncommon, for example, to find claimants in compensation cases valuing their loss of amenity very highly, yet being 'satisfied' with a much smaller payment. From this it is easy to imagine the difficulties of conducting an accurate survey of preferences.

A variety of techniques has been reviewed by Whitbread (1977) comparing behavioural studies, social surveys and experimental studies. The first aim to correlate changes in behaviour with changes in environment; the second seek the preferences of respondents using questionnaire surveys. Experimental studies are interesting in that they try to simulate an environmental change (e.g. traffic or aircraft noise) and to observe the behavioural response. Whitbread reports an experiment conducted by the Transport and Road Research Laboratory in which an indestructible tape recorder which operated automatically was placed in the homes of respondents, and emitted loud aircraft and traffic noises. After a few days the respondents were asked if the experiment could continue and offered money if they were unwilling. The 'bribe' was never actually paid but was used only to estimate the disutility of noise.

The third approach to elucidating attitudes is via conflict analysis – an area of research which geographers have not yet developed but which promises to be fruitful. At its most simple this involves no more than examining who the parties to a conflict are and what they want, or what ideals they express, in relation to a particular issue. From this it is possible to clarify the conflict according to its characteristics (Deutsch 1973). This is discussed in Chapter 9, but the basic objective is to define the 'veracity' of the conflict; that is, whether the parties are actually concerned with the same thing, whether the object of conflict is a surrogate for something else, and so on. Assuming that the conflict is a 'true' one it is then possible to subject it to analytical procedures with the aim of uncovering the 'trade-offs' that the different parties are prepared to make. Out of all this may come a statement of group preferences in relation to what may be complex issues.

An example of the use of conflict analysis in the context of an urban transportation plan has been documented by Saaty (1978). The conflict that was analysed involved a proposal to build a subway system, a system of highways and a bridge across the Potomac River in Washington, DC in 1969. Several parties were involved and there was no sign of compromise. Saaty's approach was to rank preferences as far as possible, and to subject them to a quantitative analysis: Table 3.2 illustrates the situation as it was in 1969. The analysis then moved on to an examination of the possible compromises that might exist by postulating the options that were available to each group: they could be for or against various components of the project and, in the case of inner-city

Table 3.2 Elements in the analysis of conflict: transport planning in Washington, DC (Saaty 1978)

A. Components of the plan
3 Sisters (3S) bridge
North Central (NC) freeway
Subway
Air rights housing on NC freeway
Potomac River freeway
Center Leg freeway
South Leg freeway
Industrial highway
East Leg freeway
North Leg freeway

B. Parties in the conflict

PORS	Public Officials for Roads (backed by certain Congressmen interested in highway construction)
ICRS	Inner-City Residents
CONS	Conservationist Interests
DBIS	Downtown Business Interests
SCIS	Suburban Commuter Interests

C. Preferences of the parties

	PORS	DBIS	SCIS	ICRS	CONS
3S bridge	1	1	1		0
NC freeway	1	1	1	0	0
Subway		1	1	1	1
Air rights housing				1	
Potomac highway	1	1	1		0
Center Leg	1	1	1		
South Leg	0	1	1		0
Industrial highway	1	1		0	
East Leg	1	1	1	0	0
North Leg	1	0	1		0

'1' means favourable. '0' means unfavourable. A blank means indifferent or neutral.

residents and conservation interests, they could bring sanctions to bear on the conflict (civil disturbances and litigation, respectively). In this particular example the conflicts were indeed profound, especially among the three main parties (the public officials, the inner-city residents and the conservationists) and there was still an impasse two years later.

Summary

The actions that shape our landscapes are based on wants and ideals. These may not be shared by everyone, or even by a majority. Indeed, they may be held, initially, by only a single person; for example, a farmer who doesn't like hedgerows, or an architect like Burley Griffen with the resources and the conception to create a modern baroque city in Canberra.

When planning and decision making aim to be *responsive* to popular feeling this imposes a requirement to elucidate the wants and ideals of the population to be served or affected. How to do this, and how to express those aspirations in the form of objectives, have been the focus of this chapter. We would argue that much indeterminacy arises out of the attempt to be responsive.

The greatest problem is to come to terms with the effects of the distribution of incomes and opportunities. While there can be little doubt that these help to shape patterns of behaviour, and that these patterns would change if incomes and opportunities changed, it is another matter to predict exactly what their magnitudes and direction will be. Yet it is necessary to strive for precision, even if it is an elusive goal.

4 Resource accounts

Karl Deutsch, the political scientist, asserted that 'power consists mainly in power over nature and power over men . . . power over nature is something that men can share. Power over men is something for which men must compete' (1970).

This dichotomous view of power – which is synonymous with control over resources – is a familiar one. History is full of examples of co-operative action to control or harness the natural elements: the draining of marshes, the terracing of hillsides, the construction of dams and irrigation schemes, and so on. On a different plane, the division of labour and the pursuit of agglomeration economies represent the use of social organisation to gain power over nature. Yet the dichotomy is superficial. In particular, it is not clear that power over nature and power over men can be so easily separated. Often one may be a necessary condition to secure the other.

However, it is not the purpose at this point to debate the political ramifications of the distribution and control over resources. Rather it is to explore and discuss various aspects that have been, and may increasingly be, the concern of applied geographers. These include the identification, measurement and prediction of resources at the broad scale and, by contrast, the use and the distribution of resources at the micro-geographic – even personal – level. In this discussion it will be clear from the outset that it is not only physical or natural resources that are included but also man-made capital and human resources.

In the present chapter we focus on *accounts*. What, on the surface at least, may not seem a very exciting topic is, in fact, a basic requirement for many forms of analysis. Whether we are concerned with regional development, or urban land use, or environmental planning, there is a need for an estimate of what exists, what its capacity is, what value it has, and so on. Likewise, when drawing up a plan one needs to know what the implications are for the flow of re-

sources and their creation or diversion from other uses. Hence in the following sections we deal with four basic issues: the definition, evaluation and mapping of resources, the preparation of wealth accounts; the preparation and use of regional transaction accounts; and budgetary processes.

Definitions and measurements

'*Resources are stocks*': the importance of this statement, especially for geographic analysis, is profound. Most regional and locational theory is, in fact, about the use, creation and distribution of stocks of natural and human resources, productive capital and social infrastructure. However, most such theory also concentrates upon *activities* – on the behaviour of consumers and producers and on flows of goods, services and people. It is right that these should receive a strong emphasis because they represent the dynamic *processes* through which the different components of economic and social systems are interrelated. Yet an analysis would be incomplete that ignored the stocks that provide the necessary links between the various flows.

If information about resources is to be more than a random set of facts it has to be defined and collected purposefully. As Young (1973) has argued in relation to rural land evaluation, 'the approach must be problem-oriented, starting from a definition of aims and proceeding into whatever subject matter is necessary for their accomplishment'. Being purposeful means discriminating not only among different types and uses of resources, but also among the ways in which resources can be measured and valued. Young illustrates this (Table 4.1) in terms of rural land classification, but his argument extends to all resources (see Douglas 1973). For each purpose for which land

Table 4.1 Classification of purposes and stages in rural land evaluation (Young 1973)

Purpose	Qualitative evaluation	Quantitative evaluation	Economic evaluation

Agriculture, arable	Suitability for:
	   individual crops;
	   annuals, tree crops;
	   arable use
		Crop yields, under defined management levels	Annual income (gross net) from crop production; present worth of annual values (using discounted cash flow); cost–benefit analysis of investment

Irrigation	Suitability of land for irrigated agriculture	Crop yields, with and without irrigation	Annual income, with and without irrigation, in relation to costs of irrigation: cost–benefit analysis of investment
Engineering	Suitability for roads and other engineering works	Parameters relevant to road construction and maintenance	Costs of road construction and maintenance: benefits
Recreation	Suitability for recreation: high-intensity, low-intensity uses	Density and frequency of acceptable use	Economic evaluation of predicted recreational use
Multi-purpose evaluation	Relative suitabilities for different types of land use	Ordering of relative suitabilities in accordance with defined criteria	Ordering of suitabilities for different purposes in accordance with assigned economic values

can be used it can be defined and evaluated either qualitatively or quantitatively or in terms of economic significance.

For the time being we shall retain Young's distinction between qualitative and quantitative assessment even though, in reality, it is very difficult to define a clear boundary between them. Young regards them as separate stages in the process of survey and evaluation, corresponding respectively with initial reconnaissance (qualitative) prior to detailed study of physical and technological factors affecting capacity or productivity, which then can feed into economic evaluation. However, qualitative information does not ignore numerical analysis, although it may rely more heavily on binary (presence–absence) and ordinal (ranked) data.

Young's classification raises two issues. Firstly, it is not always possible to exercise a free choice among modes of measurement and evaluation without incurring substantial costs. Economic evaluation, in particular, requires heavy data inputs which may be not only difficult to collect but also quite unstable. The value of a crop is not independent of market prices; these, in turn, might be affected by volatile international markets. Quantitative evaluation also makes large demands on data and it may also be subjected to sudden changes rising from technological innovation. Thus in circumstances where data-gathering skills and capacities are limited, there may be little

option to do other than a qualitative evaluation. Secondly, even if there is a degree of choice it may still be preferable to opt for a more 'simple' form of evaluation, partly for the reasons just given. That is, if the valuation is subject to the vagaries of international markets, or the disruption of technological change, then it may be preferable to use a qualitative approach. The same may be true if items which possess cultural value cannot be expressed in economic terms. Another alternative is to combine elements from each of the three modes. Examples where this has been done are found in land-capability maps where the aim is to provide an indicator of potential agricultural use. In Britain this has entailed the use of soil and site factors that might affect the range of crops, their yield and costs of production under a moderately high level of management (Burnham and McRae 1974).

A further feature of this classification is that quite different results may be obtained, depending on which mode is used. Barkham (1973) illustrates this, using recreation-carrying capacity as an example. 'Carrying capacity' can be defined in a number of different ways, but the clearest contrasts come from ecological and economic definitions. The former is stated by Barkham as 'the maximum pressure of land use that can be accommodated without physical degradation': the latter (economic) definition he takes from O'Riordan (1968) – 'the maximum aggregate user satisfaction while taking into account the views of the people already there'. The latter is a Paretian welfare definition, and is by no means the only economic criterion that could be used. Barkham's own preference in practical situations is to take whichever of the definitions would impose the lesser pressure upon an area.

Waller (1970) presents a quite different view, arguing that environmental measurements should be translated into monetary values which (in theory, at least) are then amenable to cost–benefit analysis from which it would be possible to define environmental standards. Though this might be superficially attractive to some economists, it is by no means a simple task to carry out properly. Not only do present values have to be taken into account, but future ones as well. The latter may then have to be discounted to make them comparable with present values – which raises problems about choice of discount rate. Then one also may need to take into account possible environmental and economic uncertainties, and the value of environmental phenomena to different people. These may not be insurmountable issues and, indeed, economic criteria have to be taken into account, especially in the preparation and evaluation of environmental impact statements (EIS) (now required by law in a number of countries before major projects can proceed). Most EIS have turned out be weak simply because they have not been able to deal with monetary values in a comprehensive manner (Spry 1975).

The basis of the last few paragraphs is that resource inventories can be conducted in a number of different ways and with varying degrees of complexity. Some of these are illustrated below.

The simplest and more familiar are basic inventories registering only what exists at the time of the survey. A land-use map falls into this category. By itself it involves no judgement about the objects that it describes. (Hence it is not a part of Young's classification in Table 4.1.) This does not mean that no decisions are needed in collecting, coding and presenting such a survey. Indeed, the classification of information in land-use mapping can present a significant challenge which, over recent years has been met by increasing the number of categories and giving more attention to the meaning of their description. The greatest problems come in the treatment of urban land. In the first land-use survey of Britain this was described in a cavalier fashion: it was lumped together with waste. More recent classifications have been orientated to the needs of planning departments and have discriminated among different users by sector: residential, industrial, commercial, and so on. But this still has shortcomings, as Dickinson and Shaw (1977) have pointed out, in that it tends to ignore or to subsume other important attributes of land such as ownership and use. It is not altogether surprising, therefore, that there should be differing estimates of land under various kinds of use.

At higher levels of complexity one moves into qualitative and quantitative evaluation and mapping. (In the following discussion Young's distinction is dropped.) The primary purpose of evaluation is to classify land according to its *suitability* for, or *capability* under, different uses on a sustained long-term basis. However, we can also extend the discussion to other purposes such as hazardousness, potential degradation and aesthetic quality.

In each of these purposes the same basic set of requirements is met. Firstly, evaluation requires *diagnostic criteria*, or variables, which provide a basis for assessing suitability. These may be characteristics inherent in the land (e.g. its soil structure and quality, drainage, slope, stability) or an external influence upon the land (e.g. climate, location with respect to markets). Unfortunately, a problem that commonly arises, especially in terrain classification, is that of correlating land types with other variables (Ollier, 1977). Secondly, evaluation requires *comparison* between land units. Ultimately, 'suitability' has to be disaggregated to 'degrees of suitability', which means placing different land units on a meaningful scale. Thirdly, we may also need to introduce an element of *prediction* to take account of variations in productivity under different technical conditions.

Qualitative and quantitative evaluation and mapping procedures have undergone many experiments for different purposes, at different scales. Suitability for agricultural use has remained the most important area of study but environmental impact assessment has also inspired imaginative research.

An early example of mapping to define agricultural suitability is presented in Fig. 4.1, which describes rainfall reliability in East Africa (Glover *et al.* 1954, in Morgan 1975). The isohyets represent the thresholds necessary to support different crops: maize (750 mm), sorghum (500 mm). Below 500 mm per year makes an area suitable only for ranching or herding unless it can be irrigated. Probability of failure of rainfall to achieve a threshold in more than one year in five makes an area unsuitable for the relevant crop. These are rule-of-thumb

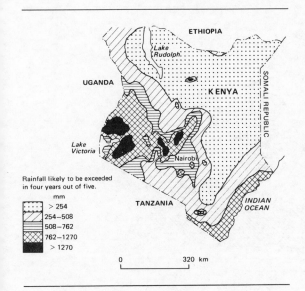

Rainfall likely to be exceeded
in four years out of five.

	mm
	> 254
	254–508
	508–762
	762–1270
	> 1270

0   320 km

Fig. 4.1 Rainfall reliability: rainfall likely to be exceeded in four years out of five (After Glover *et al.* 1954, in Morgan, 1945)

criteria, applied on a grand scale, but they corresponded roughly with experience and enabled an assessment of agricultural potential at the time.

A more detailed assessment, reported by Brinkman and Young (1976) describes the suitability of land for mechanised agriculture in Brazil. In this it was necessary to take account of the structure of agriculture as well as physical conditions. The former was divided into land-use categories (Table 4.2) ranging from capitalist, high-technology farming to primitive tree-cropping. Physical limitations on mechanisation were related to slope, stoniness and shallowness of soil, drainage, soil structure, micro-

relief (such as ant hills and gullies). These were used to define degrees of limitation (strong, moderate, slight, none) which were then used in conjunction with the categories of land use to define suitability for mechanisation. The important point to emphasise is that this study was careful to avoid associating suitability with physical factors alone.

A quite different purpose is contained in the next example which concerns a world-wide study of desertification, described by Mabbutt (1978). The impoverishment of ecosystems and the human tragedies that accompany this (as in the Sahel during the droughts of 1968–73) prompted the United Nations to ask for the preparation of a world map to delineate areas undergoing, or at risk of, desertification (Fig. 4.2). The resultant map describes the interplay of three sets of factors giving rise to the problem: bio-climatic stress, inherent vulnerability of the land and its ecosystems, and the pressure of human use. As can be seen from the extract in Fig. 4.2, the threat of desertification is at its highest beyond the margins of the full deserts – that is, where human pressure is critical. In conjunction with information on population, production and climatic fluctuations, this map contributes an important diagnostic element for future planning.

These assessments of resources have in common the application of pre-selected criteria to determine the feasibility of activities in particular areas. The same approach has been used in land-use planning in the form of filter (or sieve) mapping, where the objective is to define areas within a region which are most appropriate for the location of new development (Forbes 1969; Duffield and Coppock 1975; McHarg 1969). The technique is illustrated in Fig. 4.3, which extracts part of Forbes's work in

Table 4.2 Abstract of factors limiting the use of agricultural machinery in Brazil (after Brinkman and Young 1976)

Category	Slope (%)	Stoniness (%)	Rockiness (%)	Drainage	Depth
I Tractor efficiency > 90%	< 8				
II Tractor efficiency 60–90%	(a) < 8	0.05–1.0	2–10		shallow
	(b) < 8				
	(c) < 8			low permeability	
	(d) 8–20				
III Tractor efficiency < 60%	(a) < 8	} similar to (b) and (c) above			
	(b) < 8				
	(c) 8–20	1–15	10–25		shallow
	(d) 20–40			erosion rills	
IV Hand tools only	(a) 20–40	15–20	25–70		shallow
	(b) 40–70			erosion gullies	
V Unsuitable for implements	(a) 40–70	>40	>70		shallow
	(b) > 70				

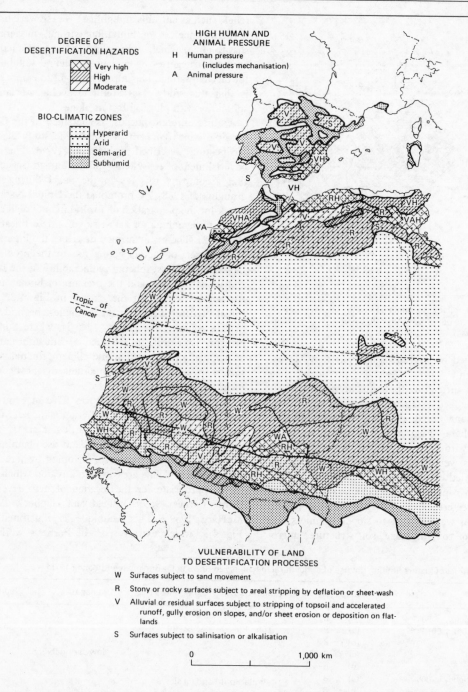

DEGREE OF
DESERTIFICATION HAZARDS

- Very high
- High
- Moderate

BIO-CLIMATIC ZONES

- Hyperarid
- Arid
- Semi-arid
- Subhumid

HIGH HUMAN AND
ANIMAL PRESSURE

H   Human pressure
        (includes mechanisation)
A   Animal pressure

VULNERABILITY OF LAND
TO DESERTIFICATION PROCESSES

W   Surfaces subject to sand movement
R   Stony or rocky surfaces subject to areal stripping by deflation or sheet-wash
V   Alluvial or residual surfaces subject to stripping of topsoil and accelerated
        runoff, gully erosion on slopes, and/or sheet erosion or deposition on flat-
        lands
S   Surfaces subject to salinisation or alkalisation

0   1,000 km

Fig. 4.2 Part of the *World Map of Desertification* by FAO, UNESCO and WMO (1977), depicting degree of desertification hazard
as a product of bioclimate, human and animal pressure, and vulnerability of land to desertification processes. Very high
hazard results from the combination of human pressure and land vulnerability

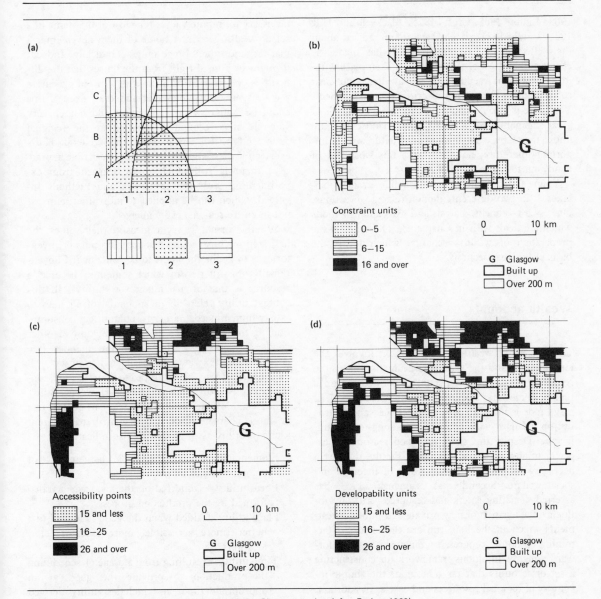

Fig. 4.3 Filter mapping for urban development in the Glasgow region (after Forbes 1969)

west central Scotland. Given a set of diagnostic variables relating to physical constraints and assessibility, Forbes presents a series of maps, out of which a composite analysis is made of land that is 'suitable' for urban development. By implication, the outcome is a representation of development costs – but without monetary values. It is also clearly dependent on the choice of variables, or filters, used in the exercise.

Finally, the last example concerns the most intangible of resources: the aesthetic quality of land-scapes. Probably, also, this has been the least satisfactorily handled, if only because diagnostic variables are the most difficult to define. Aesthetic quality is something that is holistic (one perceives it in a *gestalt* manner) whereas most approaches to the problem have sought to define individual components which contribute to the beauty of a landscape. Also, as Raymond Williams (1973) has eloquently argued, attitudes to landscape are far from static over time or homogeneous among people. Hence it is not surprising that the many experiments in this field (Fines

1968; Linton 1968; Crofts 1975; Blacksell and Gilg 1975; Liddle 1976) should have doubts raised about their effectiveness. Appleton (1975) has made the criticism that most of this work has taken place within a theoretical vacuum: there have been no clear rules to guide the setting of criteria for assessment. Yet, there is also no doubt that aesthetic quality is of great moment, particularly when landscapes are under threat by ugly, utilitarian structures. It is important, therefore, not to dismiss this work out of hand, but to try to develop it.

Young's third category – economic evaluation – takes us in two different, though related, directions. One is cost–benefit analysis and associated methods which are dealt with in Chapter 8. The other is the preparation of wealth accounts, which forms the basis of the next section.

Wealth accounts

The concept of wealth is elusive for it involves the identification and the economic evaluation of all the resources available to a person or located within a region. Sometimes a resource lies undiscovered or its magnitude is unrealised: sometimes the value that is put on it fails to take account of its long-term significance. Despite this there are good reasons for experimenting with wealth accounts. Firstly, they are a necessary supplement to the study of incomes and social welfare. Secondly, as Czamanski (1973) has asserted, 'existing assets and resources critically influence the location of new private and public investments which in the long run are the factors determining economic progress'. Czamanski has put forward a highly ambitious framework for constructing a set of accounts that far transcends the simple inventory of regional resources and assets. In addition, his discussion is accompanied by experimental work carried out on the economy of Nova Scotia, some of which is illustrated in Tables 4.3 and 4.4. Note that in these accounts the concept of wealth includes not only reproducible assets but also non-reproducible assets, financial capital, natural resources and human resources.

Czamanski's approach is to calculate monetary values for each resource and to present these under three headings: (a) type of asset; (b) ownership of the asset; and (c) its location. Immediately this raises the questions of whether such accounts can be constructed with any meaningful precision. Clearly it is necessary to impute values for most of the items in a set of wealth accounts. Some of them never enter a market, and others do so only sporadically. Indeed, it might be argued right from the outset that the difficulties of constructing a full set of accurate accounts are insurmountable and that, even if this could be achieved, the outcome would not be very valuable since economic planning at the regional scale tends to be partial rather than general. However, while these arguments may be irresistible in practice, there is still much to be learned from experimenting with the principles and methods involved, which can be applied to individual items, if not to entire regional economies.

Whether or not we try to measure their values, the first step in constructing resource accounts or inventories is to define the items to be contained. Conventionally, the criterion by which something becomes a resource is that of 'usefulness' or 'utility'. In this context utility refers to the potential of an item to satisfy human needs, either through direct consumption or through its use in the production of other goods and services. This definition includes free goods (sunshine, water, fresh air, and so on) as well as those that can be appropriated for use or consumption by individuals or groups. However, utility is not a monolithic concept: an item can possess one or more characteristics which contribute different kinds of utility as the following list indicates.

Elementary utility – the desirable characteristics of a good while in its natural state

Form utility – conferred on the good when transformed by the production process

Place utility – added when the good is moved to a better (more accessible, more efficient, etc.) location

Time utility – resulting from storage or scheduling of production to provide the good at an appropriate time; also includes utility derived from conservation of resources

Intangible utility – derived from information on the quality of the good (i.e. one may derive pleasure, and value, from an object without using it)

Quantity utility – resulting from the bridging of a gap between supply and demand (partly analogous with scale economies)

Assortment utility – refers to the presence of a range of goods in close proximity, enabling comparison or simultaneous satisfaction of a range of wants (partly analogous with external or linkage economies)

Table 4.3 Wealth accounts: types and controls of wealth in Nova Scotia, 1961 (Czamanski 1973)

Control over wealth	(in millions of dollars) Natural resources	Man-made capital	Human resources	total
Non-farm households	302.2	1,001.0	16,198.0	17,501.2
Unincorporated farms and farm households	78.1	111.8	.557.8	747.7
Business corporations	105.5	581.9		687.4
Unincorporated businesses	29.5	128.7		158.2
Non-profit organisations	4.0	125.2		129.2
Local and provincial governments	749.1	154.1		903.2
Federal government	19.6	195.0		214.6
Non-Canadian foreign units	69.8	142.0		211.8
Total	1,357.8	2,439.7	16,755.8	20,553.3

Table 4.4 Regional wealth per capita in Nova Scotia, 1961 (in dollars) (Czamanski 1973)

Location	Natural resources*	Forest reserves	Farm land and capital	Other land and capital	Human resources	Total
Halifax County	401.8	82.4	35.9	5,053.5	30,809.0	36,382.6
Cape Breton County	204.6	73.0	38.8	2,570.2	23,058.8	25,945.4
Rest of Nova Scotia	2,057.8	384.4	343.4	2,901.7	17,823.9	23,511.2
Nova Scotia	1,219.9	236.4	194.8	3,501.6	22,735.1	27,887.9

* Except forest reserves and land.

The types of asset considered by Czamanski range widely across the entire spectrum of the regional economy but, for simplicity, are presented (Table 4.4) in three categories: natural resources, man-made capital, and human resources. Natural resources include 'free goods' as well as those that are more obviously appropriatable such as land, natural vegetation and planted crops, and mineral resources. Man-made capital comprises both privately and publicly owned structures and financial assets and includes productive capital, social infrastructure and residential capital. Human resources are embodied in the population of a region.

Having identified the resources, the next step is to estimate their value. Inevitably this poses difficulties that can only be resolved by making highly simplifying assumptions, even for those resources that appear with some frequency on the market. (The importance of a market in this context is that it provides a simple, though possibly inaccurate, estimate of the value of a good by identifying what people are willing to pay for it. Some of the problems of using market values were discussed earlier and will be taken up again in Chapter 8.) But, in any case, present market value is only one way of estimating the value of a resource. Two alternatives which might be more

appropriate under certain circumstances are: (a) replacement cost, which is the value of the resources necessary to re-create exactly the resource under consideration; and (b) capitalised future income yield, which is the expected stream of net benefits from the resource, expressed as the *present* value of resources needed to produce that stream. It may, for example, be preferable to use replacement cost when evaluating an asset which has low elementary or form utility but has great cultural value (e.g. a medieval church or a fine row of Victorian terrace houses). Capitalised future income yield might be more appropriate where present value does not reflect possible future scarcity, as might be the case with some environmental resources. In a perfect system each of these three measures might result in the same value for a resource, but in reality we would expect them to be different.

Another general problem in regional wealth accounts is the perennial geographical issue concerning regional boundaries. In this context the setting of boundaries may affect the perception of scarcity and abundance, particularly of goods which appear rarely or indirectly, if at all, in the market. For example, *within* a region there may be an 'over-abundance' of beautiful scenery and sunshine. In a closed economic system these would not have a high 'value' as aesthetic attributes since they are in oversupply relative to demand. (Though the sunshine may help agricultural productivity: but that is a different matter.) However, if neighbouring regions have a 'deficit' of scenery and sunshine then the population of the first region clearly has a pair of collective assets. Such a case was put by Ullman (1954) as a partial explanation of California's economic growth and more recently in discussions of the development of the American 'sunbelt' and 'snowbelt'.

The boundary problem may also work in the opposite direction. Scarcity of a feature in a region may lead to an overvaluation of what exists. This, in the case of environmental assets, has been put forward (Clayton 1974) as a possible cause of misguided preservation. These are general issues: more specific ones are raised when the main components of wealth are considered: natural resources, human resources, and man-made resources.

The first component is the natural resource base which Czamanski treats under four headings: minerals for national or world markets; land; biotic populations; and free goods. Each poses conceptual and methodological problems. A region's mineral resources can, as has been witnessed in a number of depressed regions, eventually become a millstone. If the resource faces a high elasticity of demand, which means it is susceptible to price changes or competition from substitutes, and if the region capitalises on its comparative advantage to the extent that other industrial developments are retarded, then the vulnerability may be great. Conversely, a newly discovered (or newly viable) resource may promote rapid growth. The problem here is in being able to put a value on such resources: unforeseen vulnerability leads to overvaluation while unforeseen viability results in undervaluation. Unfortunately, it is difficult to see a way out of this impasse. Even the estimation of physical quantities of a resource may prove to be difficult, especially in its early exploration and exploitation. This issue has been investigated in great depth by O'Dell (1976, 1978), who arrives at the conclusions, based on the oil industry, that the concept of recoverable reserves is quite volatile, that early figures tend to be underestimates, and that, in any case they tend to be a function of price, especially when the reserve is a relatively small part of the global total. Without taking further account of these factors, the resource account may present a rather static picture of a region's potential. Notwithstanding such difficulties, Czamanski's preference is to use

present values, net of extraction and transportation costs, in the valuation of mineral resources.

If these appear to be difficult issues then those surrounding the estimation and interpretation of land values might appear to be insuperable. For example, market values may be a function of population density or of accessibility (or, more likely, of both), as well as of inherent qualities of land. Thus, present value might lead to strange inter-regional comparisons in which regions with land shortages would tend to show high values, and vice versa. Here the valuations might be interpreted as reflecting past development rather than future potential. But this is only partly true, as land values also contain a major element of future expectations (Drewett 1971). This is generally shown by the phenomena of demand inflation, and more particularly by the effect of expected future oil prices on the relatively rapid increase of inner-city land values in the late 1970s in many cities around the world. For this reason Czamanski suggests that present market price is an appropriate measure.

Biotic populations also present a severe dilemma. It is possible (and common) to take a myopic view, using present market value of the existing stock. This, however, is a 'once-and-for-all' valuation of what should be treated as a renewable resource which will provide a stream of future income. If exploitation of a resource is carried out competitively by many separate producers without regulation, then it is in the interests of each to extract as much as possible in the short term. The result was summarised by Hardin (1968) in an evocative phrase: 'the tragedy of the commons'. It is also exemplified in the actions of most fishing nations until recent years. Partly because the world's oceans were freely accessible and because fish were seen as a 'free' good, the interest of each fleet was served by extraction rather than conservation. The alternative is to use a valuation which gives status to the preservation of ecological balance and the continuing stream of income available through managed extraction.

The concept of a 'free' good is a difficult one to fit into the framework of economic evaluation. 'Free' only means that no one is excluded from the use or the benefits of the good. But the good may be put to quite different uses – some of them incompatible – which, when implemented may mean that the good is no longer free (though it may still be a public good) throughout the region. In the case of water, for example, the range of uses may include domestic and industrial consumption, irrigation, power generation and cooling, transportation, fishing and re-creation. Putting values on a region's water supply is thus likely to defy any direct quantification, though Czamanski suggests that it might be possible in-directly, by assessing its affect on the productivity of water-using industries. However, this does not appear very satisfactory, as an industry is likely to be only one user among many and, in any case, is likely to have different production functions in different regions, making it difficult to attribute differential productivity to any single cause.

Turning to human resources, the second compo-nent of wealth, it is important to deal first with a moral question. All human lives are equally valuable. Any other belief carries terrible consequences for a society: slavery, forced migration, even genocide. Having said that, it is equally important to acknow-ledge that, for a variety of reasons, human beings acquire unequal attributes and these may be distri-buted geographically as well as by social class. It is these attributes – *acquired* physical and mental skills – that Czamanski attempts to take into account when discussing human resources.

At any given time a region's population contains a variety of abilities (which can be expressed as its occupation structure) and can be divided into homogeneous groups by age and sex. Czamanski sug-gests that these resources can be valued in two ways. One is to measure the replacement costs of their attributes. For a number of reasons this is likely to be an undervaluation. People have innate abilities that are useful even without formal training; and the skills acquired from training are likely to appreciate in value with experience. The other method is to estimate the future income flow of the population discounted to the present. For this it is necessary to 'age' the population into the future using a survi-vorship matrix in which death rates are assumed to be constant. It may also be assumed that employ-ment participation rates and relative earnings are constant, and that migration does not occur. (These can be relaxed to produce other projections if there are good grounds for doing so.) The result is an eva-luation of the productive potential of a region's population – albeit based on some very tenuous assumptions. Referring back to Table 4.3 it will be recalled that this is by far the largest single compo-nent of a region's wealth. If it is true that human resources are also the most important factor in a re-gion's well-being and development, then it can also be asserted that this is the most crucial diagnostic variable in regional analysis. But, as may be readily

understood, it is not easy to relate the result of this evaluation to a region's potential for development. If the ratio of human assets to other resources is high (i.e. there is underemployment) then one might ex-pect either one of two contrasting alternatives. Either labour will move out of the region; or capital will move in.

Man-made resources, the third component of wealth, can be classified in two different ways. One corresponds to ownership and comprises producer-capital, consumer-capital, infrastructure and, pos-sibly, defence. The other reflects the attribute of the resource and consists of reproducible and non-reproducible capital, intangibles and financial assets. Values for most of these items, whichever classifica-tion is used, are more readily estimated than for natural or human resources, since they are more closely aligned with market conditions. There are, however, different accounting procedures that might be adopted (book value, market value, capitalised net income, and so on), each of which may give a different value. Two issues have been important to geographical analyses of regional change. One con-cerns the valuation of housing and infrastructure, both of them immobile resources which account for substantial proportions of regional assets. The argu-ment has been put forward that regional policy might be most efficiently geared to creating or di-verting industrial capital to areas where population decline (through migration) is outstripping the con-traction in housing and infrastructure stock. The case has been reviewed by Cameron (1974), who sug-gests that a proper valuation of these assets, dis-counted to allow for their declining value rather than simply looking at their present apparent abundance, tends to show that their value declines at least as fast as the population of the depressed region. The second issue concerns the valuation of intangibles (such as a social institution) and non-reproducible assets (such as historic buildings, works of art or antiques). Clearly these influence the quality of life in a community. But, as Czamanski assets, there is no consensus about how to value them.

Before leaving the discussion of resource or wealth accounts, there are two further comments that can be made. Firstly, it has been apparent throughout this section that while identifying assets may be re-latively simple, the task of putting values on them is enormous. The fact that Czamanski was able to pre-sent such values should not mask their tentativeness. They are no more than intelligent estimates, open to substantial questioning and criticism. In response to this we can make two points. One is to consider the alternatives. Simple inventories or, (as described in the previous section) enumeration without evaluation, are likely to be unsatisfactory just because they do not go far enough. The other is that it is not necessary to follow Czamanski and attempt the evaluation of every resource. Although there is some virtue in trying to estimate a region's wealth, it is equally useful to concentrate on individual resources and inter-regional comparisons. The principles – and the difficulties – outlined above are just as relevant to this task.

Secondly, in all of this discussion we have followed Czamanski and concentrated on resources and assets. Nothing has been said about *liabilities*. As a consequence, the assertion that 'existing assets and resources critically influence... economic progress' must be regarded as only partly valid. It is not enough to state that the valuation of a resource is *net* of liabilities, or that the value put upon it reflects the difficulty of extracting or transporting it, and so on. This is clearly unsatisfactory, since the net value, stated without any qualification, may also reflect its small size or poor quality. In addition, a region may also possess attributes which have a negative effect on development but which may be overlooked if one only deals with resources and assets. What appears to be required, therefore, is a more penetrating review of liabilities – or limitations on development.

As an initial step we can define four types of liability. One, financial debts, is already incorporated in Czamanski's accounts. The others are: natural hazards; degraded, derelict or obsolete resources; and locational handicaps. Each of these may need to be overcome before a resource can be developed, and even if it can be there is certain to be an additional cost of development.

Transaction accounts

The previous section focused on resources as stocks whose value arises from their present or future utility. Now we turn to their counterpart, the flows that are generated by and augment (or deplete) resources. In making this transition we also enter the area that attracts the greater volume of theoretical and empirical research. This may reflect a predilection for process rather than form, but whatever the cause there can be little doubt that flows (the transmission of economic change, the movement of goods, people

and information, the transfer of energy between ecological niches, the disposal of wastes, and so on) present the greater intellectual challenge. Stocks, existing at one moment in time, present a static concept: flows introduce the dynamic dimension. They also introduce a greater degree of indeterminacy. Whereas stocks, to a large extent, are fixed and definable, flows have magnitudes, directions and destinations which are often difficult to predict.

Stocks and flows are fundamentally related, with flows forming the links or transmissions between stocks. This has long formed the basis of national income accounting whose intricacies can, with a little licence, be reduced to a simple scheme like that shown in Fig. 4.4. Indeed, the definition of national income is that it is the flow of payments to and from households and organisations, less that flow which corresponds to the consumption of capital (resources) used in producing that income. In Fig. 4.4 each of the boxes represents a stock of resources and the connecting lines are monetary transactions. However, these monetary transactions are, for the most part, the corollary of physical movements: trade in goods and services, transmission of information, the journey of labour from home to work, and so on. At the same time many of these physical movements are influenced by macro-economic conditions such as growth rates in GNP and disposable income, technological change, monetary and fiscal policy and regional policy.

In this section we shall concentrate on accounts, the basis of which is transactions in goods and services between industries and regions. The three examples discussed (economic base, shift and share, and input–output) have all been used in the analysis of regional development.

In order to draw up a meaningful set of accounts one needs to have a system in which all categories of transactions are unequivocable, and also a common unit of measurement. The system does not have to be closed, but if there are external transactions they need to be clearly identified. For this reason one finds good national financial accounts in almost every country but increasingly weak ones as one moves down the scale and, in consequence, increasing reliance on estimates and on the use of 'proxy' variables to describe regional transactions. This is not necessarily a bad thing: often there is more interest at the regional level in the movement of goods or the employment generated by trade than in financial flows. However, none of these measures provides a consistent, homogeneous unit of account. Conse-

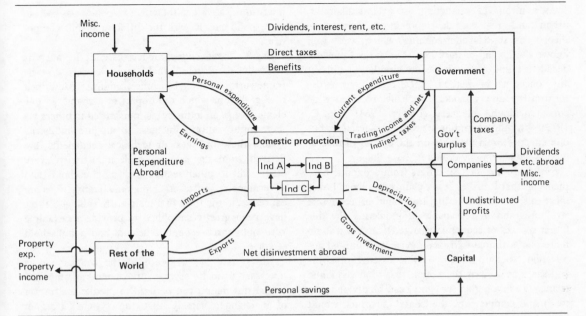

Fig. 4.4 The circular flow of income in an open economy

quently it is usually difficult to translate physical measures into monetary ones, and vice versa. The chief objective of transaction accounts is to produce a description of an urban or regional economy which can also be used to explain and predict development. In recent years the techniques of urban and regional accounts have become increasingly disaggregated and complex, but the conceptual basis has remained unchanged. Passing through economic-base theory, shift-and-share analysis, and input–output modelling, we find a broad accordance with the framework of national income accounting described in Fig. 4.4. At the regional scale this is expressed in terms of internal relationship among industries and households and external linkages with suppliers and consumers in other regions.

In its primitive form *economic-base theory* simplifies these relationships so that a regional economy is presented as just two sectors: an exporting (or basic) sector, whose sales to other regions provide the impetus for the domestic (or non-basic) sector (North 1955). Thus, one could summarise a regional economy, as:

$$E = B + S \text{ where } E \text{ is Gross Regional Output,}$$
$$B \text{ is exports, and}$$
$$S \text{ is domestic production}$$

Moreover, if the ratio between B and S was constant

one could predict the impact on S and E of a given change in B. The unsatisfactory nature of this model was quickly pointed out. Firstly, regions were never so simple. Hardly any regions live without some imports, so the income generated by the basic sector is not all retained with a region: some of it leaks out of the region. Hence, Tiebout (1956) raised the criticism that the above equation needed to be expanded to take account of this factor. Secondly, it is an oversimplification to treat B alone as the engine of change. To achieve the level of productivity necessary to generate a competitive export sector it may be necessary to have a highly efficient domestic sector (Blumenfeld 1955). Similarly, investment in the domestic sector may create extra demand for the output of the basic sector, giving it the conditions for scale economies.

Clearly this simplified, highly aggregated approach is no more than a useful initial description of a regional economy. Its application, moreover, is hindered by lack of regional income and product statistics and most analyses have had to fall back on the use of employment statistics using either location-quotient (Mattila and Thompson 1955) or minimum-requirements (Ullman and Dacey 1960) techniques to identify industries which are 'over-represented' in a region and, hence, by implication are export industries.

As a means of predicting, or even explaining, urban and regional development economic-base theory has serious shortcomings which were well known by the end of the 1950s. Yet in the following decade it found its way back into circulation without any conceptual or methodological refinement in the form of the Lowry model, which has received wide currency among planners (Lowry 1964; Lee C. 1973). Why this should have occurred is unclear, but one effect is to cast doubts on the output of macro-models which employ economic-base theory.

Shift-and-share analysis differs from economic-base theory in that it makes fewer claims on theory but is more disaggregative in its treatment of economic structures and more capable of indicating the different sources of economic change. Even so, it is no more than a simple statistical device with limited application (Richardson 1978). The essence of shift-and-share analysis is the notion that total economic change (*T*) in a region (or a city) can be divided into two major components: the 'share' (*S*) which would have occurred if the region had been following the natural trend, and a 'shift' away from that trend The latter can be further subdivided into a 'proportional' shift (*P*) which is due to the industrial structure of the region, and a 'differential' shift (*D*) which is due to individual industries changing at a different rate from their national counterparts. Thus,

$$T = S + D + P$$

Like economic-base theory, the results derived from shift-and-share analysis depend upon the definition of industries, the delineation of boundaries and the selections of starting and finishing dates. In an analysis of economic change in Merseyside, Buck (1970) experimented with two industrial classifications, one with 24 categories and the other with over 100. In the former the structural component accounted for none of the region's growth; in the latter it contributed one-third of it.

Notwithstanding these problems, shift-and-share analyses can be a useful device for the analysis of past changes – provided that one can explain why the various components performed as they did. If one cannot make this explanation the results have little meaning and the technique has no value for predictive purposes. Its use in description has covered a variety of situations including the general analysis of long-term regional change (Perloff *et al.* 1960; Stilwell 1974), short-term cyclical fluctuations in unemployment (Sant 1973), and the impact of regional policy (Moore and Rhodes 1974). The latter was an

interesting case since the effect of policy was seen to correspond closely with the differential-shift component.

Lastly, *input–output* models extend the analysis of regional change much more closely to the framework of national income accounting (described in Fig. 4.4) in which the impact of a change in one element (e.g. an industry) is transmitted to others via their input–output linkages. In an inter-industrial system these linkages, or technical coefficients, are derived from the sales of each industry to every other, and to other sectors such as households and governments, and to other regions. Transactions are expressed in the form of a matrix. In Table 4.5 these have been greatly simplified to provide an example with only three economic sectors and a household sector.

Unless one has evidence to the contrary, it is necessary to make several fundamental assumptions before this matrix can be used to predict the impact of a change; namely that the system starts in equilibrium, that technical coefficients and production functions are stable, that industries are homogeneous, and that the system is working at full capacity. Thereafter, it is a simple matter to trace the impact of a change in the output of one sector on its demands for the output of other sectors. This begins with an initial impact (round 1) and is followed by subsequent impacts, as each industry transmits its increased output to other industries, until the system returns to a new equilibrium level of total output. The ratio between the addition to total output and the initial injection of demand is the *multiplier*. In Table 4.5 it can be seen that by the fifth round, $3.294 m. worth of increased output had resulted from the initial extra demand of $1 m. Further rounds would have added to the total and the multiplier is something over 3.294.

These values are all expressed in monetary units, which is the only means of conducting an input–output analysis. However, in regional and urban development we would probably be more interested in employment and population changes, and subsequent demands for housing, infrastructure, services and land use. To translate monetary impacts into physical units requires assumptions (or evidence) about the stability of relationships between monetary and non-monetary measures similar to those made above.

Our hypothetical model ignored the government sector and inter-regional transactions. Both have great importance in full regional input–output analy-

Table 4.5 A hypothetical input–output analysis

(A) The general form of an input–output table

		Consuming Sectors 1,	2	...	n	Total C	Total X
	1	x_{11}	x_{12}	...	x_{1n}	C_1	X_1
	2	x_{21}	x_{22}	...	x_{2n}	C_2	X_2
Producing	.	.	.		.	.	.
	.	.	.		.	.	.
sectors	.	.	.		.	.	.
	n	x_{n1}	x_{n2}	...	x_{nn}	C_n	X_n
	V	V_1	V_2	...	V_n	$\Sigma V = \Sigma C$	
Total	Y	Y_1	Y_2	...	Y_n		

(B) Direct inputs per dollar of output in a hypothetical four-sector economic system.

Producing sectors	Consuming sectors Primary	Secondary	Tertiary	Households
Primary	0.3	0.5	0.1	0.2
Secondary	0.2	0.2	0.4	0.4
Tertiary	0.1		0.3	
Households	0.4	0.3	0.2	0.4
Total	1.0	1.0	1.0	1.0

(C) Round-by-round input requirements ($1,000) for an increase of $1m. output in the primary sector.

	Round 1	Round 2	Round 3	Round 4	Round 5	Total (1–5)
Primary	300	230	213	174	134	1,051
Secondary	200	220	168	125	90	803
Tertiary	100	60	24	9	4	197
Households	400	280	230	184	149	1,243
Total	1,000	790	635	492	377	3,294

ses. The government sector is often a major supplier of inputs and also a large consumer of goods and services. Moreover, it often contains a significant discretionary component and can alter programmes of expenditure which, at the regional level, can have important impacts. When this occurs, input–output analysis can be used to predict the magnitude and incidence of the change. However, a corollary is that possible long-term variations in government spending create a source of uncertainty which reduce the applicability of input–output analysis beyond the immediate future.

The importance of inter-regional transactions has already been noted in the discussion of economic-base theory. Indeed, at the regional level there may be as many linkages outside the system as there are within it. Herein lies a major dilemma for regional analysis, namely how to estimate these 'leakages' from the system when making forecasts of the impact of a change. In some cases the impact of growth in an industry is resolved quite easily. If the region is self-sufficient in an industry whose transport costs are high enough to discourage its import, then one can predict that the impact on that industry will be contained in the region and, possibly, that the industry will grow. Slightly less straightforward is the case where an industry is not represented at all in a region but whose output is nevertheless consumed in the region. Here the impact may fall entirely outside the region *or* it may be that the industry is induced to locate in the region. Finally, there are the cases where there are alternative sources of supply inside and outside the region. Here one has a situation that is indeterminate unless one knows the competitive strengths and the capacities of the various sources.

The corollary of estimating the magnitude of external leakages when an internal change occurs is the estimation of the impact of a change in one region, via its exports, upon another region. For this to be done requires a complete set of inter-regional, as well as inter-industrial, accounts. At this point one is asking for more than is provided in national income statistics.

Inevitably one is usually left with 'second-best' in regional transaction accounts, being forced to work with crude estimates and inappropriate variables (Allen 1970; Gordon 1973). This being the case, there is ample justification for major research efforts aimed at refining what is available. Two directions have been followed. One is to attack the problem directly, experimenting with economic impact models in different situations, concentrating on particular changes or industrial developments. This has been done in a variety of cases including, for example, the multiplier effects of tourist development (Archer and Owen 1971; Archer 1976). The other direction has been to concentrate on 'leakages'. A general approach was adopted by Moseley (1973b) in an attempt to identify how much of the effect of expansion programmes was retained within growth centres, and where the leakages were directed to. It was from this work that the 'trickle-down' effect (see below, p. 63) was brought into question. Another example is found in the work of Lever (1974) who examined several industries in central Scotland in order to estimate the multiplier effect that each would be likely to generate if it underwent expansion. Interestingly, he found that in some cases expansion in central Scotland would in fact have a greater impact elsewhere.

While the purpose of an economic input–output analysis is generally to determine the impact of a change on the level of a region's income, it should not be overlooked that, at the same time, the analysis is dealing with the distribution and use of resources. Indeed, rather than the conventional matrix, shown in Table 4.5, we could conceive of one in which the sectors comprised various natural, human and man-made resources. This is something of a fantasy, but there are ways of extending conventional input–output analysis beyond its common use for economic projection. For example, Kneese (1977) illustrates its use in analysing and predicting environmental pollution (Fig. 4.5). The extension consists of allocating rows and columns in the input–output matrix to the production and reduction, respectively, of 'residuals' or pollutants. Kneese suggests that the results may be interpreted in monetary or physical terms, but since pollutants have no market value it is difficult to see how a monetary interpretation can be made except by first analysing the physical magnitudes. A major experimental study using this approach in the USA endeavoured to project the impact on residual management problems arising from economic

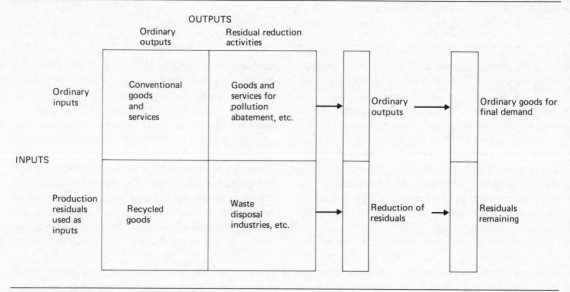

Fig. 4.5 An input–output model for environmental pollution (after Kneese 1977)

growth, population increase, technological change and different pollution policies. Of course, it can be argued that a national economy, especially that of the USA, is too large a base for such an analysis: pollution management is better done on a regional scale. Also, pollution control is not just a matter for 'residual-reducing' industries but can be managed by emission controls applied in the production sectors. Nor does an input–output model look at what happens once the pollutants enter the environment. The input–output concept does no more than provide an entry to the problem, by associating the distribution of resources with the pattern of their linkages.

Resource budgets

Lastly, we turn to a topic that is unfamiliar in geography, but which is relevant to a discussion about the creation and use of resources, namely, budgeting. Perhaps the conventional image of budgets is that they are the province of treasurers and accountants. However, 'budgeting' is really no more than a synonym for making decisions about resources.

Budgetary systems have two broad components, a current account and a capital account. The first consists of flows of funds associated with ongoing activities – payments and receipts for work done. The second comprises expenditure for the acquisition and maintenance of fixed assets, such as plant, machinery, buildings, and infrastructure. In both cases one is concerned with a stream of resources over time – perhaps on an annual basis, perhaps over a longer period. What a budget can do is assist in smoothing out the flow, by accommodating or avoiding bottlenecks and ensuring that a decision about one resource is balanced by decisions on others.

As everyone knows, budgets are not confined to central governments trying to balance the books once a year. All organisations make budgets, and in many aspects it is the 'intermediate' ones – public agencies and government departments, and local authorities – that present some of the most interesting problems, for they deal with allocation of resources that impinge closely or directly upon the welfare of people. For these agencies the question of 'where' to spend becomes a critical one, almost as important as that of 'what' and 'how' to spend. It is quite clear that much of what is being considered under the title of 'welfare

geography' (Smith 1977) would gain in applicability if it gave greater attention to budgetary processes and budgetary reform.

Perhaps a reason why geographers (and others) have tended to avoid the topic is their limited perception of what budgeting entails. Part of the problem arises from the conventional reliance of organisations on *control* budgeting, which is concerned with the regulation of expenditures ('balancing the books'), and *performance* budgeting, which deals with the efficient management of resources. Both are essential aspects of budgeting, even in a welfare-oriented society. However, both tend to be supportive of existing distributions: control budgeting because it tends to be incremental, and performance budgeting because it seeks higher net returns which, for many things, means following the market.

In contrast, recent years have seen increasing interest in a third budgetary form known as *Programme Planning and Budgeting Systems* (PPBS). This had its origins in large private enterprises, such as the motor companies in the USA, before being taken up by the American military forces in the 1920s. Thereafter, it was adopted by public services such as libraries and education departments and, only recently, by some local governments (Cutt 1978). In all of these one characteristic is dominant, namely that the organisations concerned are ones which can identify sets of objectives that they want to meet, and towards which they can design appropriate programmes. In addition, these are organisations that have a reasonably assured view of their future resources, which is a feature that is essential for the implementation of programmes. As Wildavsky (1975) has indicated there is a wide variety of budgetary types, related to wealth, the predictability of economic change, and political stability (Fig. 4.6).

PPBS requires at least a fair predictability of revenue and, preferably, a high level of revenue, though this is not essential, and there is no reason why programme budgeting should not be applied in less developed countries. However, wide economic fluctuations (even in richer countries), low political stability and weak administrative capacity all militate against PPBS. This is something that all concerned with normative, prescriptive research should bear firmly in mind. Policies are only as good as the ability to implement them and that may depend on an assured flow of resources.

Wildavsky's typology was originally developed to make international comparisons but with a little modification it can be applied to regional or depart-

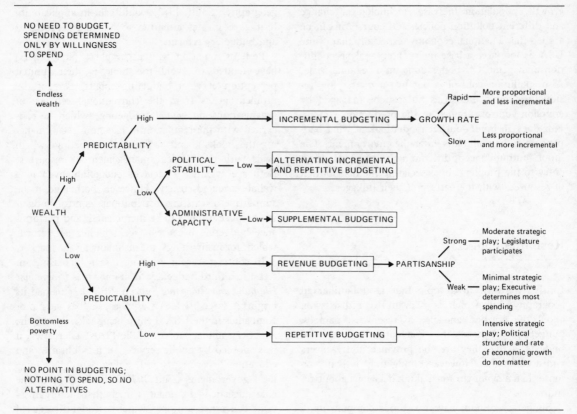

Fig. 4.6 Types of budgetary process (after Wildavsky 1975)

mental systems. As anyone involved with higher education since the early 1960s can testify, there has been a shift from incremental and proportional budgeting in recent years. The same applies to many other areas of public finance for urban and regional development.

The features of PPBS can be partly depicted by showing how it relates to traditional 'line-item' budgeting. In the latter, expenditures are listed under separate operations or items without much regard of what those items are trying to achieve. Under PPBS the same expenditure is allocated to programmes whose aim is to pursue particular sets of objectives.

However, setting up programmes creates much greater demands on the decision-making process, including a close integration with a number of practical techniques, as the following list of procedures indicates. The establishment of objectives alone may require a major analysis of social or economic indicators in order to define needs or wants. Thereafter the necessary steps might comprise:

the definition of a number of alternative programmes that might feasibly achieve each of the specified objectives;

evaluation of the different programmes leading to the selection of one (evaluation techniques are discussed in Chapter 8);

identification of major issues (e.g. resource requirements, resource bottlenecks) that might have to be overcome at various stages in the programme;

the design of a budget cycle enabling appropriations to be made as required; and creation of monitoring and evaluation facilities to assess whether objectives programmes and budgets are working out consistently, which sets demands upon information systems.

Examples of PPBS in action cover a wide range of sectors and scales. One, related to national planning for urban development was prepared by the Department of Urban and Regional Development (DURD) (1976) in Australia. Although it is a federal country,

Australian urban investment is partly funded and co-ordinated by the Commonwealth government, often through partnership with state governments. In the early 1970s it became accepted in DURD that a set of programmes should be defined with the following objectives:

1 to ensure high environmental standards in new urban development;
2 to prevent or reverse deterioration of older urban areas and raise the standard of public services;
3 to offer alternative urban environments to those found in the state capitals and other major industrial cities (where over two-thirds of Australians live); and
4 to promote regionalisation.

Relating these objectives to actual problems in the form of a number of programmes, DURD was able to identify the resource implications for the public sector (Table 4.6). Thus, the first objective – high quality of urban development – was to be implemented through the operation of land commissions in programmes of land acquisition and servicing, and through sewerage and water-supply programmes. The fourth objective – promoting regional co-operation – inspired no more than a little assistance for inter-governmental (Commonwealth, state and local) communication. In sum, these programmes implied a rapidly growing volume of expenditure from which other, apparently unconsidered, implica-

tions might be drawn. The build-up of growth centres (apart from Canberra), for example, was seen mainly as an exercise in land acquisition and assistance for development works. There was little discussion of the means whereby the population and employment targets would be achieved and what these meant for longterm budgets. Nevertheless, the DURD exercise was a useful first attempt to review national development in terms of programmes and resources rather than as a set of piecemeal expenditure through many disparate agencies each with its own objectives and accountability.

Summary

From the preparation of a basic inventory of resources to their prediction and budgeting, there is a theme which has to be observed. This is the need for consistency between the structure of accounts and the nature of their usage. It is not always possible to achieve this. Usually, the limiting factor is the availability of information which may occur in a form that prevents detailed statistical or economic investigation. The consequence is that either an analysis has to be tailored to meet the potential of the data, or that one must try to bridge the gap between data and theory by the intelligent use of inference.

Table 4.6 The Australian urban and regional budget, 1974 (McMaster and Webb 1976).

URD ministry programmes: Summary
($ m.)

Programmes	1972–73	1973–74	1974–75 budget estimate	1975–76 forward est.*	1976–77 forward est.*	Objective served (see text)
Land commissions	–	8.0	56.9	165.0	185.0	1; 2
Sewerage	–	27.9	104.7	112.0	119.0	1
Urban water supply	–	–	4.4	8.5	13.5	1
Growth centres –						
Canberra[†]	77.8	104.1	140.0	166.2	187.1	3
Others	–	9.2	82.7	133.9	167.3	3
Area improvement	–	7.4	14.1	21.3	24.3	2
Urban rehabilitation	–	5.3	16.8	10.5	10.5	2
Regional organisations assistance	–	–	0.3	0.4	0.6	4
National estate (heritage)	0.1	0.8	8.0	15.0	20.0	2
General administration n.e.c.	1.4	4.4	5.7	7.6	7.6	–
Total URD ministry programmes	79.3	167.0	433.7	640.4	734.7	

* Excluding imputed advances for capitalised interest.
[†] Including outlays of the National Capital Development Commission on national works, and Australian Government offices located in the Australian Capital Territory.

5 Circulation and distribution

In the preceding chapter a description was given of the circular flow of income in an economy. This was used as a basis for discussing regional accounts in which emphasis was placed on transactions in goods and services. However, as was foreshadowed in that discussion, the notion of circular flow has much wider importance for it also forms the basis of explanations of economic change and the distribution of wealth and welfare. A change in one sector, for example, is transmitted through other sectors to produce a multiplier effect on incomes and employment.

The objective of the present chapter is to examine more fully the characteristics of circulation in a geographic context. To begin with, this is done without any particular reference to social or economic impediments that might inhibit the free circulation of goods, people or money. Note, however, that 'free circulation' is not synonymous with any sense of equality or equity. Although the circular-flow model is conceived as being oriented towards an equilibrium state, this may or may not be associated with a tendency towards equal distributions. Equilibrium may reflect initial opportunities and advantages that are themselves unequal. The first section, therefore, describes circulation patterns and systems that have been observed, emphasising the integration of different scales of spatial interaction. This is followed by a brief review of two major alternative regional-development theories. Both are based on the circular-flow model but the differences that exist between them are crucial for the formulation of regional policies. Thereafter, we break with the assumption that circulation is unimpeded, by giving attention to discontinuities and disjunctions that occur in economic landscapes and which may contribute to patterns of inequality. Some of these are deliberate, and are related to the way in which systems are managed. Others arise from the distribution of communal and personal resources. Lastly, the circular-flow concept also provides a vehicle for analysing the impact of social and economic change brought about by locational and distributional policies.

Systems within systems?

Although most circular-flow models are represented in terms of simple aggregates, in reality circular flow is like a complex system of broad currents and minor eddies in a turbulent ocean. To understand properly any part of the system we need to understand the system as a whole, and vice versa. Herein lies a universal intellectual dilemma: namely, achieving an appropriate balance between generality and uniqueness in explanation. Too coarse a mesh may mean that important forces are overlooked, while concentration on detail may result in wider social forces being left out of the explanation. The alternative to seeking the elusive 'best' scale of analysis is to adopt a loose hierarchical approach in which each level of analysis focuses on those parts of the circular flow which are most relevant at that scale.

This has been attempted by Richardson (1973, 1978) in a synthesis of regional and urban development theory. His approach is to relate theory to patterns which, he asserts, are generally observable and which he encapsulates in the phrase 'decentralised, concentrated, dispersion'. The apparent contradiction in terms arises because it refers to the three broad scales of spatial analysis: inter-regional, regional and urban (or metropolitan). Figure 5.1 describes what is meant. At the inter-regional level, economic growth begins by being polarised in one region but is later transmitted to other regions (though, in keeping with cumulative causation theory, not necessarily to the same level as the leading region). Within a region a similar process occurs but in the form of an evolving settlement system and increasing concentration in a limited number of

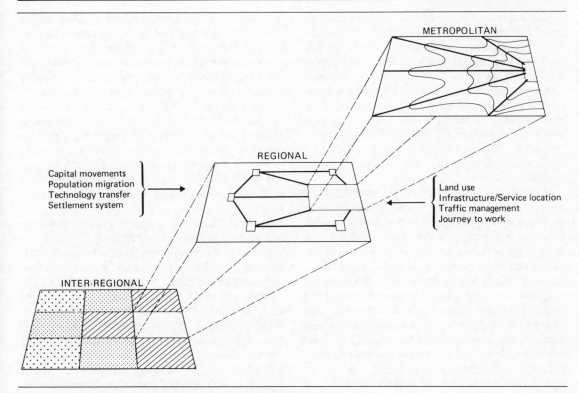

METROPOLITAN

REGIONAL

Capital movements
Population migration
Technology transfer
Settlement system

Land use
Infrastructure/Service location
Traffic management
Journey to work

INTER-REGIONAL

Fig. 5.1 Concentrated, dispersed, decentralisation: a schematic view

urban areas. At the urban scale, expansion follows a decentralised path through successive suburban and exurban developments. Each scale is associated with particular characteristics (also shown in Fig. 5.1). However, it has been argued (Bourne 1975) that regional and urban problems and policies are (and should be) increasingly interdependent.

Rather than try to define a discrete set of flows that occur at each scale, a broad dichotomy can be drawn between 'daily' and 'occasional' interaction systems. The first category recognises patterns of regular, frequent and short-lasting movements, broadly conforming to Richardson's regional and metropolitan or urban scales. The second refers to flows or movements which, in the lives of individuals, occur infrequently, perhaps at irregular intervals, and which are intended to have a long-term or even permanent duration. In this case the focus is upon regional and inter-regional flows. Thus in simplified form we have one system of flows between workplaces, residences and services, and another which comprises the regional interactions of labour, capital and technology

Clearly there are overlaps between the two categories: for example, some people migrate over short distances within suburbs and other commute over long distances between regions. Nevertheless, it is useful to retain the dichotomy even in these cases since it permits local moves to be put into a broader context. Thus, for example, local housing markets can be examined in the light of broad regional movements of population and jobs. This is illustrated in Berry's (1980) study which is directed to the question why, in a time of general decentralisation, selected inner areas in certain US cities (rather than other areas in other cities) have been the subject of substantial private residental renewal. Though incomplete, his explanation includes the locational trends of headquarters activities of American firms, and the subsequent centralisation of administrative and professional jobs.

While local flows occur in the context of wider ones, it is also true the latter are affected by the former. Thus the growth potential of a region (which, to be realised, may involve attracting flows of labour and capital from other regions) is likely to be significantly influenced by its internal structure and patterns of interaction. Chief among these are opportunities for agglomeration economies which are fostered by a fine network of interactions, and an apparent greater propensity for innovation associated with larger urban centres (Pred 1966) which is facilitated by their size and variety of industries and technical facilities. By the same token, actions to promote the development of a city (e.g. by a new transit system, or the renewal of shopping areas) have ramifications throughout a region and, possibly, beyond it.

Although it is not the only factor determining the nature and direction of circulation, there can be little doubt of the importance of capital within the system. It is this that underpins the characteristics of the three scales in Richardson's scheme. A 'leading' region becomes so because it achieves capital/labour ratios sufficiently high to accelerate its development; a regional settlement system represents the nodal and hierarchical distribution of capital; and urban morphology and land use are the outcome of investment decisions. In other words, what geographers have conventionally treated as location decisions within a framework of location theory should generally be seen as capital formation within the broader context of macro-economic theory.

It can be seen that the circular-flow model has two aspects. One is *generative*, the other *distributive*. The generative aspect refers to aggregate changes brought about by changes in any one component. The process was partly described in Chapter 4 in the discussion of input–output analysis, in which a change in demand for the output of one industry was transmitted through others to effect a change in total product. For the broader context described in Fig. 5.1, these inter-industrial linkages are but a single component in an economic landscape: their role is no greater than that of any other component. Aggregate changes are generated through the linkages between all the parts of the daily and occasional interaction systems.

At the same time it is almost a truism to say that circulation has distributive effects. This has always formed a plank in the marxist theory of value to explain the transference of wealth from labour to capital. The argument is that labour power is sold like any other commodity, at a price determined by the labour time necessary to produce it. But labour power is able to produce a value greater than its own value, because a worker can work more hours than are necessary to earn enough to keep himself alive.

The product of this surplus labour is called surplus value and is appropriated by the employer who has the legal right to the full use of the labour power by paying the worker the labour-time value of that power. The inequality is exacerbated when capital is scarce in relation to labour. However, the marxist model rarely occurs so obviously in practice. Labour can also be scarce, especially when it carries special skills: it can also increase its bargaining power by collective action. In addition, most governments temper the upward redistribution of wealth by policies which distribute it back towards labour through progressive taxation and the provision of public goods.

But even if capital is limited by external regulation and internal motives and inadequacies, it nevertheless seeks profits and prefers them to be larger rather than smaller. Hence it is likely to seek or create opportunities for higher rates of return and its circulation will reflect this. Inter-regional capital flows take the form of private savings in some regions being transferred into investment in others; and firms follow locational strategies which lower their operating costs or expand their markets. In urban economies there are variations in profitability among suburbs as well as between different forms of investment, so that commercial and residential development occur selectively and, usually, in response to perceived and anticipated effective demand rather than any concept of social need. We should expect this to be true particularly of private capital, but it may also be true of public capital seeking to satisfy corporate objectives.

Thus although many economists have rejected the marxist argument that distributive effects occur through the *control* of capital, it can still be asserted that a similar outcome (though less identifiable with class structures) occurs as a result of *access* to capital in the form of job opportunities, infrastructure and services. Propinquity can be as important as ownership.

Movement and interaction

The generative and distributive features of the circular-flow model have two important implications for the analysis of movement and interaction. The first is that their patterns are transitional between one state and another. What we observe at one moment represents one frame in a moving picture in which each pattern shows some influence from its predecessor and, in turn, is helping to shape its successor. The second implication is that transition may involve *stress*. Movement and interaction patterns contain two components. One reflects existing spatial structures and opportunities; the other reflects a search for new opportunities and the creation of new structures. Each may conflict with the others, which is one reason why analyses of spatial interaction tend to show correlations that are substantially less than perfect. The problem for applied geography, therefore, is not only to describe existing patterns but also to separate what is long established from what is new, to distinguish the substantive from the ephemeral, and to seek in the present the clues to the patterns that might emerge in the future.

Movement and interaction studies have a long history though formal, quantitative analyses are more modern in origin. Many of the principles upon which the latter are founded were elucidated well before societies were mobilised by industrial and technological progress, as a reading of Daniel Defoe's description of the effects of interaction and accessibility on urban development in the eighteenth century clearly shows. While contemporary analysts have the benefits of massive data banks and high-speed computers with which to test and experiment on more rigorously defined models, the basic elements used to explain interaction patterns have changed little over the years. Indeed, a criticism of some modern work is that it tends to give insufficient weight to *all* of the principles underlying movement and interaction.

Four sets of principles provide the basis for movement and interaction studies:

moves or contacts are initiated by utilitarian motives;
patterns of movement reflect the complementary and competitive character of origins and destinations;
the volume of movement is related to the size of origins and destinations and their intervening distances; and
movement and interaction are externally controlled (i.e. they occur within systems of modes, paths and land use that are determined by private and public organisational structures).

Of these, the second and third have had the highest profile in empirical studies and hypothesis testing while the first has tended to be assumed without further investigation and the fourth has received little attention. In other words, the emphasis has been on explaining observed patterns rather than examining the deep-rooted causes of those patterns or changes in them. The implication for forecasting movement and interaction is profound, for it means that one can only expect to forecast accurately if the (dis)utility of movement and the external structural controls remain constant. If the first of these is unknown, the forecast is likely to be quantitatively inaccurate; that is, the forecast will anticipate more or less movement than will actually occur because it has under- or overestimated the real cost and demand for movement. If the controlling system is also unknown, however, there is the additional likelihood of a forecast being directionally inaccurate because it lacks a sound basis of estimation.

The importance of utilitarian motives has always been recognised. Thus, Ravenstein's (1885, 1889) explanation of migration noted the trend from worse-off to better-off regions. More recent formulations have treated the same thing quantitatively by specifying the economic factors likely to stimulate a migratory move. Commonly this is put in terms of the difference between discounted future benefits arising from a move and the costs of making the move (Willis 1974). These costs may include a component for loss of personal enjoyment caused by moving from a familiar place. To be fully applicable, however, the principle of utility needs to be subjected to a marginal analysis. In effect one should ask what the marginal propensity (or probability) to migrate is at different levels of total benefit and cost, and also in relation to individual items in the equation. Similar questions apply to all forms of travel behaviour and also to industrial movement.

Complementarity and competition influence the directions of interaction. The former is most obviously the case in movement between residential and industrial or commercial areas, and in international trade. That is, one place has a certain requirement and the other has the means of filling it. However, complementarity is insufficient by itself to explain patterns of movement or interaction. An origin may have many separate potential complementarities but interact with only a few of them. The reason may lie partly in patterns of accessibility; other things being equal, interactions reflect the disutility (or cost) imposed by distance from potential destinations. Accessibility has traditionally been seen by geographers as a major factor contributing, passively, to the competitive attraction of places. However, active competition occurs among places, with each promoting its own attractions such that, given a choice of complementarities, an interaction will favour one rather than another despite their relative geographic attributes. Competition for investment (in the form of industrial relocation) exemplifies the refusal of communities to accept a 'natural order' (Camina 1974).

Apart from these aspects there is another side to competition which is important in studies of interaction: namely, the bargaining power of the participants. This has particular significance where interactions also involve transactions, for the power of the parties influences the benefits that each receives as well as the pattern of interaction. To illustrate this we can again use the example of the hypothetical ice-cream seller described in Chapter 3. A single seller with monopoly power would logically locate centrally and charge a price which maximised profits. The interaction pattern would be centralised, and consumers would receive minimal consumer surplus. On the other hand, a highly competitive market with evenly spaced sellers would result in a dispersed pattern of movements, lower prices and greater consumer surplus. In reality, inequality in bargaining power comes in a number of forms, all of them involving some degree of monopoly or monopsony. Pure monopoly, where a firm is the sole supplier of a good with no close substitute is rare, but collusion among firms to control output and prices has similar effects and many countries have passed legislation to limit it. On the other hand, elements of monopoly have been fostered officially in most countries by protective tariff duties and quotas, and by the establishment of marketing boards for primary products. Lastly, monopolistic power is vested in the supply of many public goods (e.g. health services, utilities, transport, and so on) which, while different from private monopoly in their objectives, may still result in a loss of consumer surplus if the policy of the service is to operate in a centralised manner.

The third principle, referring to the size and distance apart of origins and destinations, has had its most common expression in the gravity model (Haggett 1965; Olsson 1965). Undoubtedly there is, over a very wide range of interactions, a relationship between observed patterns and these variables. Distance (expressed in miles, or cost of travel, or time required to go from place to place) is a disutility which discourages movement and inhibits interaction among places. Size of origin influences the volume of potential movers or sources of messages; and size of destination influences the attractive force for many types of move. Like distance, size also needs to be expressed in appropriate measures. For example, in the migration of working-age people it might be suitable to use employment, while recreation travel might use disposable income at origins and some measure of recreation capacity and opportunity at the destinations.

The gravity model forms part of the spatial paradigm that has dominated modern geography and its use has extended to all of the interactions and scales described in Fig. 5.1. It precise mode of use has differed, with some studies using it deterministically and others stochastically. In the former category its components are used as independent variables against which to correlate an observed movement pattern. In the latter its use is to simulate a pattern which can then be correlated with actual patterns. It has also obtained wide currency in planning studies to forecast movement patterns and demand for transportation. Most simply, it appears in trip-distribution models where, having forecast the number of trips to be generated by each zone (based on its land use), these are then allocated among destinations, thus (Cantilli 1978):

$$T_{ij} = a_i b_j G_i A_j f(D_{ij})$$

where T_{ij} is the number of trips from zone i to zone j;

D_{ij} is the distance (or travel time) from zone i to zone j;

G_i is the number of trips generated in zone i; and

A_j is the number of trips attracted to zone j.

and a_i and b_j are values formed by successive iterations of the model.

More sophisticated analysis is afforded by the entropy model and by modal split and trip assignment models (Wilson 1974; Batty 1976) which aim to forecast the division of transport into types of vehicle and to allocate traffic to the most probable routes taken by travellers. However, these are basically extensions of the gravity model. It has also been used in conjunction with economic-base theory in the Lowry model in order to forecast regional distributions of traffic, residential growth and employment growth (Lowry 1964; Lee C. 1973).

Yet, despite its wide use the gravity model continues to raise some disquiet chiefly, it can be asserted, because it exists in a vacuum. It can 'explain' a distribution, but it cannot explain why that distribution originated and evolved in that way rather than another or what the controls are on interaction patterns. Nor is it very accurate in forecasting marginal changes in distributions when new settlement structures or modes of transport are involved, or when people's tastes are changing. This has been argued by Olsson (1978) in a critique of his own earlier work in which he bases his case on the failure of Stockholm's traffic planners to allow for changing preferences when they prepared a new public transit system for the city's developing suburbs.

Above all, this imposes a requirement for close attention to the fourth principle expressed earlier; namely, *patterns* of movement and interaction are externally controlled. That is, while an individual may have control over his personal preferences, he does not control the system through which those preferences are satisfied. Rather, he has to operate within a distribution of settlement and land use, where decisions about transport modes and route capacities (and prices) are taken by others, and in which societal functions are highly structured. The exercise of personal choice in movement and interaction is deeply constrained.

Apart from their being inherited from the past, settlement patterns and land use are significantly influenced by the extent and nature of public planning. The density of development, its location, and the relationship of one land use to others can all be affected. It would be surprising to find the same outcomes where intervention is minimal and (say) where it follows the principles of indicative planning (see p. 12). In the former category the goal of a developer may or may not be served by integrating his investment with existing settlement and land-use elements. If profitability is a function of costs (a feature associated particularly with tight housing and land markets) then there is a tendency towards dispersal beyond the urban periphery. This is even stronger when the costs of servicing land are passed on to the purchaser. In contrast, an indicative planning framework, which at least would attempt to maximise net social benefit, would normally aim at a close integration between new and existing elements of settlement and land use.

A similar general distinction can be drawn in the effect on transportation systems. Routes tend to be public goods and, as such, should attract a level of expenditure which balances capacity and use, and which is allocated with a whole system in mind. The reality may not always match these principles, however, and variations in quality of systems may reinforce the patterns of accessibility to which they give rise. The other major aspect of transport systems (traffic modes) is also influenced by public policy, or lack of it. The balance between private and public modes is affected not only by investment policies towards the latter but also by pricing policies towards the former in the shape of fuel prices, licence fees and taxes on the purchase of vehicles. Since private and public modes differ in their flexibility, a system in which the former prevails is likely to exhibit a more varied interaction pattern.

Finally, patterns of interaction also reflect the segregation and structure of societal functions. At its most basic this embodies the separation of residences and workplaces enforced by land-use planning which influences the journey to work. In rapidly growing cities where residential decentralisation exceeds employment decentralisation, this can lead to rapidly increasing journey distances and travel times. A similar outcome arises when activities, especially public services, are the object of increasing economies of scales. Larger units may enable the costs on public revenue to be kept down, but at the expense of longer travel for the users of services (Black 1977).

Specialisation of social and economic functions has been shown to have a marked spatial dimension within cities and regions, and to give rise to clearly defined patterns of interaction based on industrial complementarity (Goddard 1975). However, the same thing arises from the occupational and functional complementarity which occurs within any organisation (Fig. 5.2). Broadly, functions can be divided into categories on the basis of their role in an organisation: managerial, operational, clerical, and so on. Some need face-to-face contact; others do not. In large organisations (e.g. government departments and private corporations) these functions may be divided among precisely defined sections which, with the aid of communication systems, can be geographically segregated. Thus, 'orientation' functions which need access to their counterparts in other organisations can remain centrally located, while 'programmed' functions which may need a large labour supply and cheap rents may be decentralised. The point is, however, that this is a trend made possible by changes in corporate structure and technological development. It does not follow that such a segregation of functions will take place; nor is the form it will take necessarily predetermined. But if it does occur, the effect is likely to be a major shift in patterns of movement and communication (Hamilton 1974).

Two implications arise from these comments. The first concerns the stability of interaction equations.

(a)

JOB FUNCTIONS

LINKS and FLOWS

Administrative functions

Decision-making, planning,
negotiations, search, product
development etc.

Control direction of
production, information-
processing, services to
A_1 etc.

Accounting, routine
office work, services to
A_1 and A_2 etc.

Service functions

Business services

Household services

Manufacturing functions

Processing of materials,
handling of goods,
construction, etc.

Primary functions

Agriculture, mining,
energy production, etc.

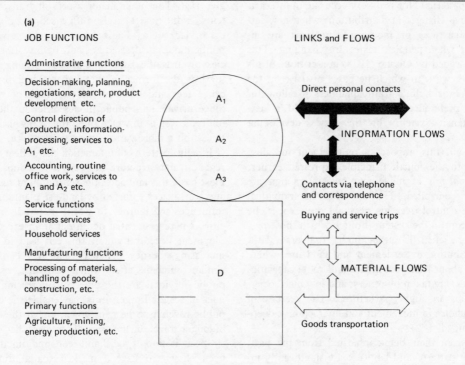

Direct personal contacts

INFORMATION FLOWS

Contacts via telephone
and correspondence

Buying and service trips

MATERIAL FLOWS

Goods transportation

(b)

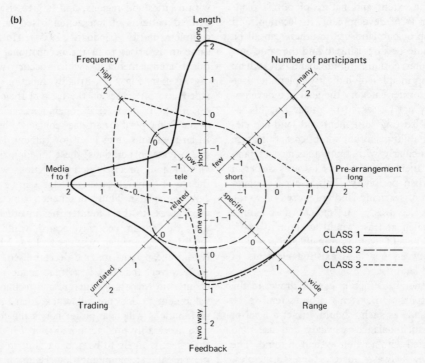

Fig. 5.2 Activity systems, job functions and interconnections (Goddard 1975)

This is not simply a question of patterns remaining constant: that can never be in a society that is continually changing. What is at issue is the relationship between interaction and the forces that influence it: whether, over time and under different circumstances, for example, the 'lapse rate' of movement in relation to distance (or cost) is the same. We already know the answer to this: it is not stable. But we know less about why it is not stable, and about how changes take place. Although we can hazard many explanations, these will not always provide a satisfactory basis for planning and forecasting. In particular, it is important to distinguish the extent to which an interaction pattern reflects personal preferences and how much it reflects inherited patterns and external forces.

The second implication is that external forces are important not only in shaping patterns of interaction but also in influencing their efficiency and equity. Unless they are instructed to the contrary, we could expect agencies (public and private) involved in development to use interaction patterns to their own advantage. The outcome may be efficient and equitable in social welfare terms, but there is no guarantee of this.

Development and disparities

Economic development is an elusive concept. At its fullest it means the improvement of the quality of life of individuals and communities; at its narrowest it is the increase of per capita income. The former is the more satisfactory definition for it encompasses things that are valued, but which may not appear in income accounts, such as security or access to free goods. Moreover, the latter includes payments for goods and services which may give a perverse kind of satisfaction. That is, they protect rather than contribute to welfare. Examples include such apparently trivial items as the purchase of burglar alarms, or insulation to keep out unwanted traffic noise, as well as the more obvious police services and defence forces.

Unfortunately, when we discuss regional economic development as a general phenomenon we tend to be thrown back on economic aggregates which can be related to per capita incomes through the circular-flow model such as investment, employment and unemployment, activity rates and expenditure on goods and services. Apart from being items for which data is available (even if it is not ideal), these aggregates do have logical interrelationships which have long been the subject of macro-economic theory.

The same cannot necessarily be said for other social indicators which would be embraced by the fuller definition of development. Health, education, the provision of cultural facilities, freedom from racial tension, and the like, all contribute to well-being but often have no obvious relationship with economic variables or with each other. Nor should we expect them to; in the process of maximising growth in the short term some of these indicators may represent an unproductive use of resources. Similarly, economic development may conflict with environmental ideals in city and country alike. Thus a danger of the narrower definition of development is that it neglects much that is of value. This is no more acceptable in post-industrial societies than it is in pre-industrial ones. In neither case is the undiluted pursuit of higher per capita income an absolute virtue.

Yet, while this is quite obvious to practically everyone, it also remains the case that regional development theory is couched in macro-economic terms. It would be easy to use this as a basis of criticism – if there were a ready alternative. However, our earlier discussion of wealth accounts indicated how difficult it would be to include every non-market and free good in a development theory, as would be required by the wider definition. Logic and practicability preclude this, and suggest that an alternative is necessary. One possibility is to follow Mishan (1967) in uncovering the costs of economic growth – whether or not they are measurable in a formal sense – and to evaluate them in comparison with the anticipated benefits of higher average per capita incomes. These costs may arise in different ways: the creation of external diseconomies (noise, congestion, pollution, etc.) which may not be costed in the production process; the loss of natural amenity; increasing social inequality; and so on. By being prepared to make these comparisons the continued concentration on macro-economic aggregates remains acceptable.

This is what we shall focus on in the remainder of this section, in which emphasis is placed upon regional economic development theory. In it we pass, first, through a discussion of the two major conflicting theories and their implications for policy making, to the problems of policy implementation and, finally, to the challenge posed by regional development theory to economic geographers.

Initial inequalities in regional incomes can arise from a number of causes: from unequal endowments of resources in relation to population; from differences in accessibility to materials and markets; or from variations in technology. In a real situation, where inequality is a prime mover for regional policy, it is essential to identify the causes, since different causes might warrant different policies. However, the development of theory has been less concerned with this than with the question of how inequalities are affected by the process of economic change.

Regional economic change has been approached by two theories which in many respects are very similar. They both focus on the same processes and are based on the same motivations, but they reach widely different interpretations and conclusions. One stresses the tendency towards *convergence* of regional incomes through equilibrating mechanisms. The other asserts that regions tend to *diverge* and that convergence, if it occurs at all, only occurs as a response to the initial divergence.

The first, *neo-classical theory*, has its roots in international economics, whose early concern was with resource use and allocation under highly simplified conditions. Although these were relaxed in order to be more realistic, the initial assumptions of neo-classical theory (profit-maximising motives, omniscient and perfectly competitive behaviour, perfect mobility of resources and factors of production, and constant unit costs) are such that the equilibrium outcome is factor-price equalisation. That is, every factor (land, labour, capital and management) receives the same return wherever it is located. However, each factor may combine differently with other factors at the various locations, so that some are capital intensive and others are labour intensive, and so on. Inequalities may occur, through discovery of a new resource, exhaustion of an old one or some other cause, but the inherent tendency towards equilibrium, brought about by perfect mobility, always forces the system back towards equality. The logic of the model is acceptable, but not the premises upon which it is built.

Cumulative causation theory, the principal alternative, contains a subtle but vital distinction. Instead of constant unit costs it recognises the existence of increasing returns to scale (and decreasing returns, as well). It also incorporates imperfections in mobility, though this is not essential to produce the effects predicted by the model. In other respects the two approaches are similar, but the outcome of cumulative causation is, as Myrdal has argued (1957), that

'the play of forces in the market normally tends to increase rather than to decrease, the inequalities between regions'. The reason is that growth in a region leads to agglomeration economies in production, an expanding local market and a larger local tax base, which in turn may set the conditions for further growth. Potentially higher profitability and wage rates attract resources from other regions (especially capital and labour) which tend to reinforce agglomeration economies and inequalities. Eventually there may be countervailing forces which encourage convergence of regional incomes. Agglomeration economies may be overtaken by congestion and decreasing returns to scale; the terms of trade between leading and lagging regions may change as the demand for the latter's products grow; and there may be industries which, in face of the rapid rise of factor costs in the leading region, find it advantageous to relocate in the lagging regions. There is evidence that all of these do occur and that convergence does follow divergence (Williamson 1965). However, it is a slow process; in the USA convergence took place many decades after the onset of divergence (Fig. 5.3) and was assisted by the pattern of federal expenditures which favoured lagging regions. The conclusion of most cumulative-causation theorists is that regional

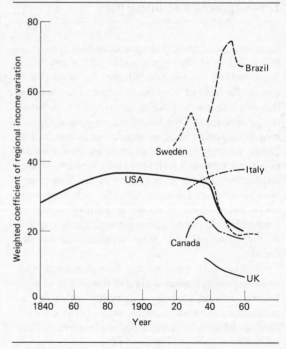

Fig. 5.3 Trends in regional disparities in selected countries

policy is needed in order to bring about effective convergence.

The reason for dwelling on these conflicting theoretical standpoints is that the decision to accept one rather than the other has far-reaching consequences for the prescription of regional policy and planning (King and Clark 1978). Modern regional policies, almost without exception, have been based on cumulative causation theory and the notion that, left to itself, a regional economic system will be dominated by the real or perceived forces of agglomeration. Hence the practice, in most countries that have regional policies, of intervening in the distribution of man-made resources (especially reproducible capital, including infrastructure) through inducements for industrial movement and the management of public investment in favour of less well-off regions (Brown and Burroughs 1977). It was not always so. Early policy in Britain, through the operation of the Labour Transference Act (1926) was based on the neo-classical idea that where there was high unemployment (or a labour surplus) the appropriate way to restore equality and equilibrium was through assisting voluntary labour migration. Following the logic of cumulative causation, this would be most likely to reinforce agglomeration economies and widen disparities.

In contrast, French regional policy in the 1960s set about not merely combating agglomerative tendencies (as the British were doing) but also establishing new agglomerative forces through the designation of *métropoles d'équilibre*. These were eight major population centres (following Paris), mostly located in regions where economic regeneration was required. The background to this policy, growth-pole theory (Perroux 1950; Boudeville 1966, 1972), is a variant of the cumulative-causation approach. Its objective is to simulate the conditions of existing areas of rapid growth by concentrating metropolitan functions (e.g. government offices, research and development activities, higher education, etc.) in the designated centres, and by focusing transport and communications linkages on them. By doing this, it was hoped, these centres would: (a) relieve the pressure on a rapidly growing, congested, metropolitan region by diverting investment from it; (b) attract migrants from declining regions who would otherwise have gone to the metropolitan region; and (c) provide an impetus for development in the growth centre's hinterland (Darwent 1969; Moseley 1974).

The growth-centre concept provides a useful example of the difficulties that surround the translation of theory into policy. Following the French initiative, the application of the concept has been discussed, and often carried out, in a number of countries ranging from Sweden to Australia, and from Ireland to Tanzania: these have been comprehensively reviewed by Moseley (1974). Unfortunately, ideas often have a habit of travelling badly and becoming confused in the exigencies of local conditions. In the mêlée to implement growth centres, their definitions became very loose, expectations were expressed which had little bearing in reality, and the conflicts likely to be raised by such a policy were often overlooked.

On the question of definitions, Moseley (1973a) produced a devastating critique in which a number of statements were juxtaposed which showed that the concept had come to mean almost anything. Places were being referred to, or designated as, growth centres which could stand virtually no chance of accumulating the conditions prescribed by Perroux and his followers: industrial linkages, size and agglomeration economies, and geographic nodality. The term was even used of villages in Newfoundland designed to centralise a couple of thousand dispersed fishing families.

Expectations derived from growth-centre theory are difficult to substantiate or to dislodge. Diverting growth from congested areas or creating an impetus for regional growth are both long-term policies which work slowly and incrementally. But some doubts have been expressed. For example, the notion gained credence that 'growth impulses trickle down to smaller places and ultimately infuse dynamism into even the most tradition-bound peripheries' (Friedmann 1966). This 'trickle-down' effect (which is almost analogous with the spread effect of cumulative causation) is an attractive idea, but little was known about what was being spread, and how the process worked – or even if it worked. It was a notion that appeared to have been adopted from diffusion theory (Hagerstrand 1967) without substantive testing. However, given the trends in aggregate consumption patterns found in many countries and the location of most growth industries in metropolitan regions and existing high-order regional centres, it could be argued that growth centres would be accompanied by strong 'trickle-up' effects, with growth impulses being transmitted up the urban hierarchy. This might be less likely when the designated centre is initially large, contains sufficient economic variety to supply incoming industries, and has well-developed links with its hinterland. But empirical work in two small towns (15,000 population) in Britain showed that 'trickle-up' was a real force which could produce perverse results (Sant and Moseley 1977).

Conflict is almost inevitable when a policy requires the diversion of resources from one place to another even if, in the long run, this results in the creation of new resources and opportunities. In the case of growth-centre development the diversion appears to involve more than just an inducement for firms to move to designated areas. It also requires a centralisation of public investment in a few selected areas: expenditure on transport systems, hospitals, higher education, and other public services is focused on these nodes. Consequently, several types of area can legitimately feel that they are being discriminated against. Provincial cities not selected as growth centres may argue that they are placed at a disadvantage, and outer metropolitan suburbs (where residential growth has outstripped social services and job opportunities) may assert a claim on these public resources. An additional stress is likely to occur when national economies are growing slowly and the diversion has to be made from a fairly static budget. Together, these factors have led to growth-centre policies being diluted. Selectivity has been balanced by general inducements for decentralisation and in several instances growth-centre targets have been reduced, if not abandoned. The latter has occurred in Australia (where the Orange–Bathurst target was reduced from 240,000 to 130,000 by the end of the century due partly to reduced national population forecasts), and in France where opinion changed to favour a stronger role for Paris in the context of European regional development.

The hiatus between theory and policy places the onus for improvement upon theoreticians rather than upon practitioners. Considerable progress has been made in partial analysis where aspects of regional development (e.g. migration and industrial movement) have been examined individually. However, general regional models have been more difficult to substantiate. Their importance might be questioned by those who see policy as a political function with a marginal input from theory, but the earlier discussion on the realism of expectations indicates why such models are necessary.

Richardson (1973) has derived an influential model which he summarises in a 'reduced form' in the following definitional equation:

$$Y = [ak + (1 - a)l]^{\alpha} + t$$

where y = growth rate of income, k = growth rate of capital, l = growth rate of labour, t = rate of technical progress, a = capital's share in income, $(1-a)$ = labour's share in income and α is an exponent reflecting agglomeration economies and diseconomies, depending on whether its value is greater than or less than one. A suggested expansion of the model is shown in Table 5.1

Interestingly, in his later work (1978) Richardson uses the same model, with one amendment, to describe neo-classical theory. The difference is that the agglomeration exponent is omitted. Strictly speaking, however, he might also have amended technological progress, much of which is oriented towards achieving economies of scale.

The challenge of this model lies in making it operational, a task which is yet to be achieved. To do so

Table 5.1 Richardson's general theory of regional economic development

(A) Definitional growth equation

$$y = [ak + (1 - a) \, l]^{\alpha} + t \qquad \alpha = 1$$

where y = growth rate of regional income, k = growth rate of capital, l = growth rate of labour, t = rate of technical progress, a = capital's share in income, $(a - a)$ labour's share in income, α = exponent, the value of which reflects increasing, constant or decreasing returns to scale according to whether α is greater than, equal to, or less than unity.

(B) Components of change
  (i) Capital stock

$$k = b_1 A + b_2 y - b_3 K - b_4 \overset{z}{CV}(K_i/\pi d_i^2) + b_5 (R - \bar{R})$$

where A = a measure of regional agglomeration economies, K = regional capital stock, $\overset{z}{CV}(K_i/\pi d^2)$ = coefficient of variation of the capital stock per unit of area in each city ($i = 1$) of the z urban centres in the region, $\bar{R}$ = rate of return on capital in the region, $\bar{R}$ = average rate of return in the rest of the inter-regional system.

  (ii) Labour supply

$$l = b_6 n = b_7 A + b_8 P + b_{14} (W - \bar{W})$$

where n = rate of natural increase of population, P = a measure of average locational preferences, W = regional wage, $\bar{W}$ = average wage in rest of system.

  (iii) Technological progress

$$t = b_{15} A + b_{16} k + b_{17} G_{n1} + b_{18} qt$$

where G_{n1} = rank of region's leading city in the national urban hierarchy, q = a measure of the degree of the region's connectivity with the rest of the economy, t = national rate of technical progress.

would require, firstly, a satisfactory definition and explanation of the independent variables in the equation and, secondly, appropriate data with which to test the model. The latter is a perennial issue in all social analysis and will not be discussed here, except to note that in this, as in all models of change, it is essential to have time-series data.

Richardson's expansion of the model is put forward as a suggestion rather than an established set of relationships and therefore can be subjected to numerous amendments. However, it does have the essential ingredient of a general model in that each of the independent variables is related to the others. The capital-stock equation (Table 5.1, equation B(i)) contains agglomeration economies, existing capital stock, rate of return, and urban concentration. An immediate point to be raised is that this treats capital homogeneously whereas we might be able to subdivide it into private financial flows, public investment and industrial movement. The labour-supply function (equation B (ii)) also shows the influence of agglomeration economies as well as natural increase, locational preferences and wage differentials. The first and last are interpreted as forces for attraction and the preference variable is seen as a reflection of a region's retentive power. Equation B (iii) outlines the technical-progress function which contains the rank of the region's leading city in the national urban hierarchy, the region's connectedness to the rest of the country (the equivalent of population potential) and the national rate of technical change. Pred (1966) has provided historical evidence for the first of these variables, while Berry (1970) has illustrated the effects of the second as well as the first. However, technical change might also be postulated to be related to capital movements, particularly through the movement of industry (Forsythe 1972). A formal discussion of this relationship is contained in Vernon's (1966) *product cycle theory*, which was initially developed to explain the international diffusion of industries as they became standardised and highly capital intensive. (Standardisation permits the use of cheap, unskilled labour.) The concept has also been applied to regional development (Norton and Rees 1979), in which it has been shown that among US regions the most rapid growth has been demonstrated by those southern and western states which are 'importing' high-technology industries, many of which underwent their embryonic development in the traditional manufacturing belt. The position is now being reached when these regions are themselves becoming innovators.

Finally, the agglomeration function needs to be carefully examined, for it contains several separate elements: business, social and household economies (and diseconomies) form three broad categories. But all three pose difficulties for identification and measurement for they require evidence that marginal costs of production are affected by size of city or region (note that size also has to be defined) and by linkages with other activities. Despite the significance accorded to agglomeration since the earliest theories of location (Weber 1909), there is little conclusive evidence on the structure of agglomeration economies at aggregated areal scales. Every change in technology brings about a change in production function. In addition, every activity has a unique production function, so that one might be enjoying scale economies at a particular city size while another is suffering from diseconomies. Not surprisingly, where 'optimum' city size, based on the cost of service provision, has been calculated the results have shown a wide dispersion of optima. Moreover, the attributes of every city and region are also different, giving rise to another source of variation.

Yet, despite all the practical problems we know that agglomeration economies and diseconomies do exist and that they are sought by private firms and public planners.

Inter alia, we may note that cost-minimisation approaches to regional planning, such as embodied in *threshold analysis* (Malisz 1969; Hughes and Kozlowski 1968; Kozlowski and Hughes 1972), are to a considerable extent based on the identification of potential economies and diseconomies of scale. In threshold analysis the aim is to compare trajectories of marginal capital costs of development in different locations recognising, at the same time, that these costs may undergo sharp shifts whenever new public facilities need to be installed before further growth can proceed (Fig. 5.4). For example, a town of 10,000 inhabitants might be able to accommodate an extra 2,000 without altering its infrastructure, at which point it will need to augment its sewerage system. Thereafter, it will reach new thresholds as the capacities of other facilities are filled: a new (or enlarged) water supply, school, central parking facilities, public transport, and so on.

What is the value of such an intricate general model as Richardson's? The fact that it is barely possible to test many parts of it, let alone the entire framework of equations, might lead to the suggestion that it is an academic luxury, with little practical value. In response there are two points in its favour.

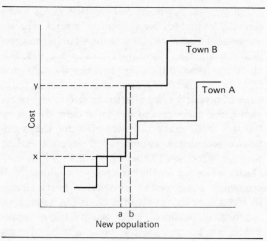

Fig. 5.4 The threshold concept in urban development: hypothetical costs of augmenting infrastructure

The first is the importance of having a general framework within which to develop partial theories and analyses. The second, more practical, value is that a general framework is necessary to provide a sound basis for identifying regional problems and for forming expectations about the effects of policies. For example, disparities might be due to any one of the independent variables in the definitional equation, or to some combination of them. Finding out which ones are responsible is a prerequisite for designing a policy. Likewise, a clear appreciation of the nature of the policy within the context of that equation assists in predicting its effects. For example, a policy which is limited to increasing capital stock in a region will certainly have some impact on per capita incomes, but not as much as if it also incorporates the realisation of agglomeration economies and technical change. Indeed, a criticism levelled against British regional policy is that it has tended to move low-technology industry to the assisted areas (Massey 1979).

Discontinuities and disjunctions

In a discussion of regional development and policy, Donnison (1974) used an evocative analogy which can sustain a more lengthy treatment. The picture that he draws is one in which a satisfying life for all the members of a community depends upon there being a full set of scalable 'ladders of opportunity' in

each place: 'We must try to ensure, in each activity, that more rewarding opportunities are within easy reach, financially, spatially and culturally. The rungs in the ladder must be placed closer together. Moreover, each ladder of opportunity must be sustained by others in neighbouring sectors of the economy.' This is not so much a plea for balanced growth in the traditional economic sense as an awareness that lack of opportunity may inhibit the full development of communities. Why bother if it gets you nowhere? Why not just move to where the grass is greener? Thus, Donnison continues, 'We must beware of *discontinuities*, where the rungs of a ladder are missing or lead nowhere, and *disjunctions*, where neighboring ladders are not available to provide mutually supporting opportunities.' Donnison illustrates his argument with British examples: Wales, where educational opportunity is rich but skilled employed opportunities are few; Scotland, where the housing ladder comprises cheap public rental accommodation and very expensive private housing for owner-occupation, and very little in between.

The cases that Donnison discusses are at a regional scale and relate to life-opportunities, but his concepts apply with at least equal force in large-city environments where questions of accessibility – financial, spatial and cultural – have an immediate day-to-day significance. Also, this scale is the scene of some major discontinuities and disjunctions. These are not just variations in accessibility based on a trade-off between travel costs and residential costs as postulated in Alonso's (1964) neo-classical, equilibrating, bid-rent model of land use. In that model people do have different opportunities available to them depending on where they live, but this is regarded as their own choice. If schools, hospitals, shopping centres and jobs are a long way from them, then (at least in theory) they are compensated by lower residential costs. Moreover, they could (with the same disposable income) make a trade-off between consuming more land with poorer access in an outer suburb and living at a higher density with better access closer to the city centre.

However, the assumption that people can make infinitely fine marginal adjustments in their consumption functions is patently not valid, for a number of reasons:

While resources (money) might be almost infinitely divisible, the commodities, services, land and accommodation that are offered for sale are not. Although it may vary in quality and price, the house, car, or holiday that one purchases is a *whole* one, involving a commitment of a block of one's resources for a certain period of time. The choice of one item may therefore preclude choice among other items.

Freedom to enter a market may be limited. A person may need to have a certain minimum level of capital funds before 'buying in' to a market even though his current income is sufficient to sustain him in it.

Freedom to enter a market may also be restricted by race or sex. Housing and labour markets have a long record of discrimination in this respect.

In the case of public goods, where there is no market, the quality and availability of items is determined by public authorities and their institutional frameworks. Choice may be limited by boundaries (e.g. between school areas) and the ability of the household or individual to meet qualifying criteria (e.g. to enter public housing).

People are not infinitely mobile. Given the means of transport at their disposal (car, cycle, foot) and the nature of their time commitments, they are limited to circumscribed 'action spaces' within which they can move. Also, given the location of different activities and services there may be clear disjunctions between what they would like to do and what they are able to do, not because they lack the money but because they lack the time.

It is not possible to examine every disjunction or discontinuity that arises because of these limiting factors. Instead we shall concentrate on two areas of inquiry that have attracted a lot of interest in recent years; namely, the operation of housing markets and the character and effects of 'time–space' budgets.

Cities contain many different interest groups and 'each grouping has an accountability pattern differing in its consequences for action' (Form 1954). Although the reality is far more complex and heterogeneous, we can think of three broad interest groups, all having different motives and attitudes towards the city: the property industry (including lending institutions), municipal government and residents. Expressed rather crudely, their objectives may be stated to be, respectively, profit, managerial efficiency and consumer surplus. Each also has a different role to play: the first two are the major suppliers of housing stock, the third are the consumers. However, the first and second differ from each other, with government sometimes supplementing the private sector and sometimes competing with it.

Public (council) housing fulfils the supplementary role, whereas assistance to home owners for the improvement of existing stock (possibly accompanied by 'gentrification') may be competitive.

The consumer's role in the housing market is only as independent as his private resources allow him to be. For the great majority of people, buying or renting a home means being placed in a client status with one or other of the suppliers of housing, or housing finance. Only if there were a universal oversupply would this not occur: but that situation is barely conceivable. It is important, therefore, to inquire how the various groups fulfil their roles.

A number of studies has followed this theme. Boddy (1976), in a study of the structure of mortgage finance in Britain, contrasts the practices of private (building society) and public (local government) lending, pointing to the different attitudes of these sources of money. There is little doubt that many building societies operate, to some degree, a policy of 'redlining' certain districts, making it difficult for their residents to borrow. It may be that they perceive the properties and residents of these areas as 'bad risks' (Williams 1976, 1978) but the outcome is a further attenuation of the disparity between rich and poor areas. It also means that the risk has to be taken up by others: the residents themselves (sometimes having to borrow from 'unofficial' sources at very high rates – especially prevalent among immigrant groups – or from local government. The conclusion reached by a number of researchers has been summarised by Duncan (1974, 1976): 'the housing system does not operate to distribute resources according to need'. This is a strong statement, particularly when applied to countries, like Britain, which have a large public housing sector, one of whose objectives is to compensate for disadvantage. But Duncan and others (Hamnett 1973; Gray 1976; Dennis 1978) have suggested the operation of an 'inverse care law' by which the management of public housing stocks and the availability of finance for house improvement have both tended to favour middle-income families. Nor should this be surprising, for municipal governments also have a set of motives and accountabilities which tend to favour security (Pahl 1970).

Although most of these studies have an overtly radical ideological viewpoint, their subject matter is an issue for general concern. Housing is a major determinant of quality of life and if there are severe discontinuities in the housing market then the ability of an individual, by his own efforts, to improve the satisfaction afforded by his housing and residential environmental is seriously restricted. It appears also that the market, and public stock management, have segregating effects. This is bad enough when the segregation is by class, but the problems are compounded when it is also by race. When segregation is reinforced by unequal access to services then one can seriously question whether an urban social system is likely to be political stable. One cannot necessarily draw a conclusive relationship between discontinuities in the housing market and civil disorders, but intuitively it is likely that the former will add to a feeling of discontent.

The second area of recent interest is 'time–space budgets', a concept developed by Hagerstrand (1970) and taken further by a number of followers (Carlstein 1977; Thrift 1977; Carlstein, Thrift and Parkes 1977). The concept is very simple and amounts to the application of common sense to everyday problems of accessibility. If a person has commitments to be at certain places at certain times of the day (or week, etc.) then, depending on the transport available to him, he has a limited spatial area for other activities (Fig. 5.5). The crucial point is as follows: what opportunities – for jobs, recreation, social services, and so on – occur within his 'time-prism', or action space? If all regular requirements are served within the time-prism, there is no problem of discontinuity or disjunction. However, there are many people for whom time imposes a real barrier or disjunction; for them, Donnison's mutually supporting ladders of opportunity are simply not accessible within the time *normally* available. Everyone has a prism of some kind, but the most restricted in scale are where we would expect to find them – among the poor, the housewives without cars, the young, and the very old.

If people were able to adjust their residential locations to maximise the content of their time-prisms and if services and other opportunities were widely dispersed, much of the problem would disappear. However, the trend in recent years has been for fewer, larger units to supply social services of many different kinds. In the provision of public goods (e.g. hospitals, schools, public transport) there can be significant trade-offs between scale economies for the producer (i.e. efficiency) and accessibility to the consumer (i.e. equity). Much depends on the operating objectives of the agency providing the service and the degree to which a government is prepared to subsidise it. In the 1960s a substantial amount of

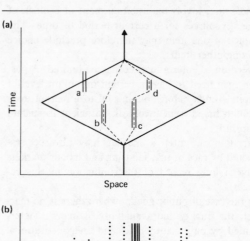

(a)

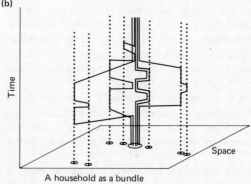

(b)

A household as a bundle

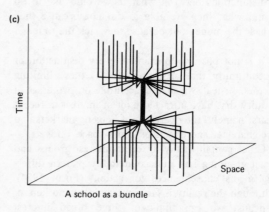

(c)

A school as a bundle

Fig. 5.5 Hypothetical time–space budgets (Carlstein 1981) (a) The 'prism' of potential movement of an individual (b) and (c) Social groupings

work was carried out on location and allocation problems posed by this issue (i.e. the location of facilities and the allocation of service areas) but most of it was based on efficiency objectives, such as minimising the total distance between scattered consumers and the central facility (Scott 1971; Massam 1975). This tended to be compounded by a prevailing preference for bigness in pursuit of scale economies – larger schools, bigger hospitals. The effect is most readily identifiable in rural regions (Moseley 1979) but it also apparent in urban areas. Nor is it only confined to the delivery of services. Many sectors of employment, especially in private and public offices, are highly centralised, while their labour markets have become more suburbanised. Many of the consequences – isolation from both services and from employment of opportunities – have been documented by Smith (1977) and there has been a shift of interest towards adapting the framework of location–allocation studies to include other objectives.

6 Uncertainty and risk

Economic and environmental systems are not only dynamic: they are also volatile and sensitive. Both are capable of dramatic fluctuations and responses to external change over short and long periods. Indeed, it might be asserted that variability and irregularity, rather than stability and regularity, are their normal conditions. With perfect knowledge (and foreknowledge) and a capacity to adjust our actions, this would pose few problems. Natural hazards could be avoided, if not averted, and economic management would overcome the booms and busts created by sudden shifts in supply and demand. But rarely do we possess anything like perfect knowledge. Rather, what we have are varying degrees of uncertainty and risk.

Uncertainty and risk mean different things. *Uncertainty* refers to lack of knowledge about the probabilities of future events. For example, uncertainty exists in the case where not only do we not know whether or not it will rain tomorrow but also we do not know the probability of it raining. *Risk* refers to situations where the probability of an event is known or can be estimated on the basis of other known conditions (e.g. the probability of a certain horse winning a race). Uncertainty breeds superstition, risk breeds insurance.

In general the applied scientist is concerned: (a) to identify uncertainties; (b) to reduce uncertainties to measurable risks; and then (c) to reduce risks, preferably to the point at which certainty takes over. Included in these is the need to expose spurious claims of certainty and the false confidence which often gives rise to the most hazardous situations both for human welfare and for ecosystems. The first and second are primarily tasks of information gathering and analysis and the third is a matter of defining actions or policies appropriate to the phenomenon. For geographers these require that they come to terms with different dimensions of environmental and social uncertainty, and with the nature of possible risk-reducing actions.

In a limited sense, any study concerned with elucidating the relationships within and between environmental and human systems helps to reduce uncertainty. In the present context, however, we are concerned with more specific aspects. One is the sources of uncertainty which are largely related to such environmental and economic characteristics as the variability, instability and irregularity of trends, events and responses. Reinforcing these sources of uncertainty, the second aspect concerns the perception of opportunities and limitations and the 'landscape' effects of ignorance, uncertainty and risk-taking. Thirdly, there is an aspect that relates closely to the subject matter of Chapter 8, the evaluation of plans and proposals in which it is often required that the uncertain effects (costs and benefits) of an action are projected into an uncertain future.

Environmental uncertainty and risk

From the viewpoint of human use, environmental uncertainty concerns the suitability of sites for particular activities. This, in turn, can be subdivided into three sets of conditions. The first is related to the parameters of the environment in its natural state: its stability, the frequency and effects of hazardous events of different magnitude, and so on. All of these affect whether and how a site may be used, and the risks incurred in using it. Secondly, it is related to the sensitivity of an environment to different forms of human intervention or development (whose own effectiveness may be influenced by the response of the environmental system within which it is placed). Finally, the physical environment provides media through which the effects of human action are transmitted to other people and places. Summarising these conditions, the first two concern the internal effects (costs and benefits) which arise from the use of an environmental resource in a particular way, while the third concerns the external effects of environmental impacts on other people.

In all three conditions the reduction of uncertainty is a function of experience and experiment. The latter includes the analysis, simulation and prediction of environmental parameters with the objective of providing information to guide human actions in an appropriate way. Examples include: early-warning systems for climatic and tectonic hazards; analysis of seasonal variability and its effects on economic activity; estimation of the magnitude and 'return-period' of extreme events and their consequent implications for built structures; and the environmental impact of various forms of interference. For experimentation to be effective it is necessary that there should be a sound basis of knowledge and understanding of environmental systems, including an awareness of what to look for. Substantial progress has been made towards this in many areas of applied geomorphology (Hails 1977) and applied climatology (Smith 1975). Thus knowledge of the properties of rocks and soils under certain geomorphic and climatological conditions can be used with confidence to assess their suitability for building different structures (Chandler 1977). Similarly, from known relationships it can be predicted that a river channel will be subjected to particular forces following a change in regimen (Schumm 1977). However, the knowledge and understanding of environmental systems is not always adequate for the task in hand and sometimes important questions are left unasked. The reason may simply be lack of experience in the proposed development. Moss (1976) asserts that this was the case in the failure to predict the geomedical effects following the construction of Lake Kariba and also in the detrimental impact of the Aswan dam on the Mediterranean sardine fishery through the reduction of nutrients being washed into the sea. In these instances experience is contingent upon human action and once the unforeseen impact has been registered it ought thereafter to influence subsequent experiments.

Also, in these cases 'experience' is something that can be readily identified in the form of a cause–effect relationship arising in clearly defined systems. However, in other cases there is a greater degree of indeterminacy about the nature of experience. This is illustrated by variability in hydrologic and climatological cycles. Most places exhibit seasonal rhythms of some kind, but these often contain wide fluctuations which may be far from rhythmic. Extremes may be not only infrequent, but also quite irregular. Even if the fluctuations are not great they may be sufficient to jeopardise the success of certain activities or impose a need for costly innovations which artificially create reliable conditions. The problem posed by environmental variability is, therefore, a two-fold one. It concerns, firstly, the ability of the analyst to provide information on the long-term variability of conditions in situations where records are insufficient to include every conceivable extreme. Secondly, it concerns the forecasting of future conditions over long, as well as short, periods (Chorley 1971).

Reliability and the return-period of extremes have been analysed in a number of different ways, but most of them focus on the *probability* of certain conditions recurring. In the example in Fig. 6.1, which describes the frequency of observed climatic conditions, probability is derived from observation. The same result could have been derived by using Gumbel graphs (Fig. 6.2), on which are plotted all the available data for a location on one axis and the return periods of the various magnitudes of the phenomenon on the other axis. The example in Fig. 6.2

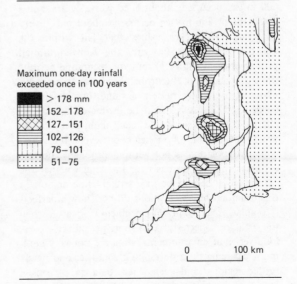

Maximum one-day rainfall
exceeded once in 100 years

- ▓ > 178 mm
- 152–178
- 127–151
- 102–126
- 76–101
- 51–75

0   100 km

Fig. 6.1 Maximum 24 hour precipitation in Wales likely to be exceeded once in 100 years (after Rodda 1967)

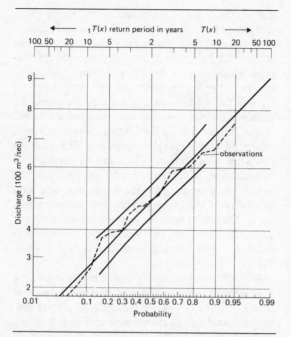

Fig. 6.2 Predicting the return period of events: a Gumbel analysis of streamflow in the Colorado River, 1922– 39 (Chorley 1971)

describes the mean annual discharges of the Colorado River at Bright Angel during 1922–39 (i.e. before the river was dammed upstream). From it one can interpret the probability of a discharge exceeding 600 m³/sec occurring about once every 4 years, and a discharge of 250 m³/sec occurring about once every 10 years. These are values that fall within the observed recordings (as with the Kenyan rainfall probabilities in Fig. 4.1), but the technique permits the extrapolation of the probability curve to give estimates of the return period of values outside the recorded set. Thus, it might be estimated that a discharge of 800 m³/sec would recur about every 25–30 years, and one of 900 m³/sec about every 100 years. This extrapolation is not without problems, however, since there is no basis for knowing whether the extrapolation should be in a straight line or not: in this example the observations do not show a perfectly regular progression. The problem is greatest when the data are collected over a short period. However, long series of environmental data are not very common, particularly in developing countries and newly settled regions; and even where data do go back a long way they often are affected by changes brought about by intensive human intervention.

Accurate forecasting requires a firm basis of understanding of the processes and relationships which underlie the variability of environmental phenomena. Furthermore, it requires not only a continuous flow of diagnostic information but also the facility to analyse it quickly enough to present a useful picture of evolving conditions to consumers of information. In many ways this is a 'cleft-stick' position since the better the data (i.e. the more abundant) the longer it takes to process it. Also, by insisting on the most up-to-date information one shortens the time available for analysis. Some of these problems can be overcome if a well-tested model is available, with continuous remote sensing of data being fed into a high-speed, large-capacity computer as has been the case with short-term weather forecasting (Smith 1975). Even so, the results tend to be qualitative: the ability to forecast how much rain, or what the precise temperature will be in a particular place at a certain time remains an elusive goal. The same applies to other areas of environmental forecasting: for example, the prediction of slope failure, either in a natural state or after modification, has yet to reach a satisfactory level of accuracy (Chandler 1977).

Although the probable accuracy of a forecast is greater over the short term than the long term, there are situations in which accurate long-range forecasts would be useful in planning the use of resources. Table 6.1 describes the uses of different forecasting horizons for weather and climate applied to a variety of economic activities. Here short term means less than 48 hours and long term means 28 days: in addition some speculative comments are provided about the potential uses of forecasting climatic change over many years. Since many economic decisions deal with the allocation and use of resources over periods of up to several decades (Table 7.3), this could be a fruitful area of investigation. (It could also, if inaccurate forecasts gained credibility, be a very expensive exercise.) As it is, short-term weather forecasting has been claimed to yield a high benefit–cost ratio (Smith 1975).

A different aspect of forecasting arises in the assessment of environmental impact. Here concern is focused on the long-term, possibly permanent, effect of projects on the quality of environments. From the preceding comments on the problems of identifying and predicting the variability of conditions it can be anticipated that this, too, is an area of some indeterminacy. Indeed, it is not uncommon to find expert opinion divided over probable impacts (Thier 1979; Sewell and Little 1973).

Methods of impact assessment vary (see

Table 6.1 Temporal horizons for weather forecasting (Mason, 1970)

Activity	48 h	1 week	1 month	Season	Climate changes
Agriculture	Day-to-day operations. Frost protection.	Timing and planning of ploughing, sowing, harvesting, hay-making, crop-spraying, fertilizer application, etc.	Timing of sowing and harvesting. Estimates of demand for fruit and vegetables.	Forward planning of crop schedules. Choice of varieties. Forecasting of crop and milk yields.	Long-term planning of land use. Breeding of new varieties. Investment decisions on buildings and machinery.
Building and construction	Day-to-day operations, avoidance of worst effects of heavy rain, frost, strong winds, etc.	Planning of operations with alternate plans to suit weather. Hiring of plant and machinery.	Planning of work schedules. Weatherproofing of buildings during construction. Hiring of plant and machinery.	Firmer estimates of completion dates, delays, etc. Weatherproofing of buildings, during construction.	Long-term planning, siting of new towns, industries, motorways. Design of buildings, structures, utilities, etc.
Electricity and gas	Hourly values, 48 h ahead of weather parameters influencing demand.	Continuously up-dated forecasts of all weather elements affecting demand for as far ahead as possible. Planning of maintenance schedules.		Forecasts of seasonal demands. Warning of abnormal spells for maintenance schedules and operational stand-by equipment.	Long-term planning, design and location of new power stations, storage and distribution systems.
Water resources	Forecasting of precipitation, evaporation, runoff, river flow. Regulation of dams and reservoirs. Irrigation requirements. Flood forecasting.			Forecasting of seasonal precipitation and droughts. Estimates of water balance, irrigation requirements.	Long-range planning and estimates of demands. Location and design of dams, reservoirs. Management of water resources.

Chapter 8) but it is possible to divide them into two broad groups: static and dynamic. Static impacts refer to those which have immediate, once-and-for-all effects: an example is the visual impact of a large building. Dynamic impacts are the more important, and refer to the modification of continuing environmental processes. Here the prediction usually employs some form of systems approach in which the proposed project is treated as an additional element in the transmission and feedback of existing environmental elements. The new element is likely to have an effect on both the rate and the direction of transmission and may thus alter the properties of the environment. For example, the extraction of water for industrial cooling changes the flow and the temperature of water downstream from the plant, which in turn affects both the properties of the water and the ecological system dependent upon it. The impact on a dynamic system may also eventually be sufficient to carry it over a *threshold* of stability. Like a dripping

tap which eventually undermines and leads to the collapse of a house, an environmental system may eventually be subjected to pressures which bring about catastrophic change. Schumm (1977, 1979) uses this concept to describe slope failure in semi-arid regions where storms or flood often trigger off unstable geomorphic conditions on valley floors which were previously oversteepened by the gradual accumulation of sediments. Once a threshold of accumulation has been passed, the trigger can lead to severe gullying which can be expensive to combat. The trigger need not be a natural event, such as a heavy storm: stability may also be jeopardised by irrigation or over-grazing.

The preceding paragraphs have focused on the problems of identifying, estimating and predicting the variability of environmental phenomena and thereby reducing uncertainty. However, this uncertainty, which we may label *positive*, is also accompanied by *perceptual* uncertainty. That is, whatever the realities of environmental variation, people may place their own valuation on existing conditions and form their own expectations about future ones. Often the positive and the perceptual coincide, especially where uncertainty has been substantially reduced through experience and research. Sometimes, however, major uncertainty, possibly combined with economic pressures, may lead to serious exposure to environmental hazards. This is conceptually different from deliberate risk-taking strategies (discussed below) where the probabilities of gain and loss and the size of rewards and penalties are measurable.

Although environmental perception has been studied for a number of purposes (e.g. aesthetic quality – Appleton, 1975 – and physical comfort – Smith, 1975) the theme of hazard adjustment presents a major challenge for geographic research on the use of resources at scales ranging from the local area to the major region. It also deals with a variety of hazards, ranging from the sudden devastation of a cyclone to the creeping damage of a drought. Many case studies have been conducted which indicate the variability of perceptions and the hiatus between perceived and actual conditions (White 1974). A leading example is contained in Saarinen's survey in the Great Plains of the perception of drought among farmers in areas which differ in the amount of precipitation received. The more humid the area, it appears, the more closely did the farmers' perceptions of drought correlate with extreme conditions: mild and moderate droughts tended to be overlooked in these areas. This finding is supported, as one might expect, in a number of studies: the greater the exposure to, and experience of, the hazard, the more accurate and discriminating the perception (Saarinen, 1976).

Perception studies are interesting in their own right, but to make the most use of them they need to be combined with other aspects of hazard research. This has been done by Kates (1971) in the form of a conceptual framework for hazard adjustment (Fig. 6.3). Here a natural hazard is defined as the outcome of interaction between a *human use system* and a *natural event system* which includes such parameters as the magnitude, frequency and duration of the event. The hazard produces *hazard effects* (i.e. loss of life and property) which in turn lead to a multi-faceted *adjustment process* which may incorporate adjustments of either the human use or, where feasible, of the natural event, or both. Or there may be no adjustment. Burton and Kates (1964) uncovered a wide variety of perceptions in response to hazards. Some people deny the very existence of the hazard;

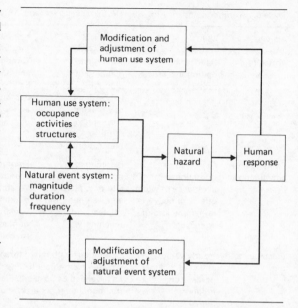

Fig. 6.3 Human adjustment to natural hazards (after Kates 1971)

some doubt its recurrence ('lightning never strikes twice in the same place'); some treat it superstitiously; and others are happy to leave action to what they perceive to be an omnipotent and benign government. Notwithstanding these eccentricities, Kates's approach is a valuable one within which to organise the management of hazard adjustment since it incorporates feasibility and welfare effects in the decision process. For example, in the response to a hazard there may be a choice between land-use regulation (and prohibition), insurance, disaster relief and environmental modification – or some combination of these. The ultimate choice requires a systematic review of the alternatives and their effects which can be carried out through Kates's framework (see Fig. 6.4).

Risk-taking can be defined as the choice of actions when the variability of environmental conditions is known (including their probabilities of occurrence) but where it is not possible to predict what the conditions will be during the period over which the action takes place. A complicating factor is that the level of risk can be affected by the amount of 'warning' that can be given. Figure 6.5 describes this in the example of flood protection for store owners. This applies to a short-term hazard requiring immediate responses, but the same principle applies to all risk-taking situations where continuous physical processes are involved: the more up-to-date the fore-

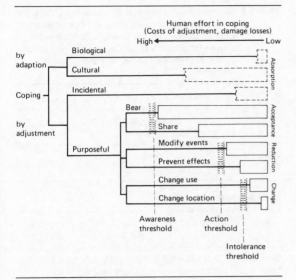

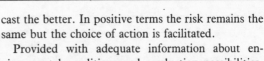

Fig. 6.4 Modes of coping with natural hazards (Burton, Kates and White 1978)

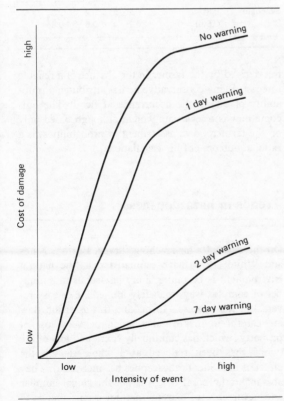

Fig. 6.5 Warning and damage

cast the better. In positive terms the risk remains the same but the choice of action is facilitated.

Provided with adequate information about environmental conditions and production possibilities, it is possible to apply *game theory* in order to select a strategy which simultaneously avoids excessive risk and yields an acceptable rate of return. This is illustrated by a simple example in Fig. 6.6, in which a farming community in Ghana is faced with a choice of cultivation strategies where wet and dry years are equally frequent (Gould 1963) but where cropping decisions have to be taken in advance of growing conditions. Some crops do well in wet years, others prefer dry. The wrong choice of crop could be serious for nutritional standards in a subsistence economy: the wrong choice in succeeding years could lead to famine. Thus the community needs to be sure that its strategy at least assures survival. The best solution is achieved by planting a mixture of maize (77.4% of available land) and hill rice (22.6%). No information is available about the quality of nutrition yielded by this strategy, but no other gives as high a *minimum* level of output. In reality, of course, this is a highly simplified example, employing the most elementary game theory. The environment is treated as a binary variable whereas different levels of wetness or dryness would be more realistic. Similarly, few societies are wholly subsistent and a strategy would be likely to be influenced by market prices.

The importance of accurate environmental forecasting is illustrated in Table 6.2 in which an hypothetical example is presented where a farmer has to decide whether to evacuate his livestock from an area endangered by flood. The effects of different conditions and choices of strategy are known. In the first case (Table 6.2(A)) the farmer is completely uncertain about the probability of a flood occurring. But he does know that while evacuation will cost him 50 per cent of his income it will at least avoid the risk of losing everything. However, without further information he has no basis for acting one way or the other. In the second case the farmer is assumed to have a reliable, objective forecast of the probability of flood: there is a 30 per cent chance that it will occur (p = 0.3) and a 70 per cent chance that it will not. Multiplying the effects of different actions by these probabilities permits the farmer to calculate the expected utility of each action. Provided he can bear the small risk of a total loss his rational strategy would be to stay put. Lastly, Table 6.2(C) repeats the analysis, but this time using subjective probabili-

(a)

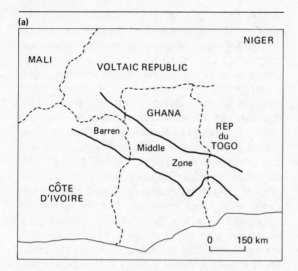

(b)

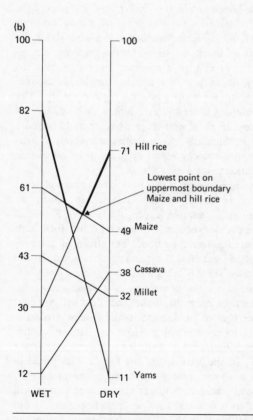

Fig. 6.6 Decision making under uncertainty: crop combinations in Ghana (Gould 1963)

Table 6.2 Objectivity, subjectivity, uncertainty and choice. (In these hypothetical cases a farmer has to decide whether to evacuate his livestock from a low-lying area endangered by flood. Each example describes a rational processes of choice under different conditions.)

A. Complete uncertainty

Action	States of nature	
	Flood	No flood
Evacuate	+2	+2
Stay	0	+4

B. Objective probability

Action	States of nature		Expected utility
	Flood	No flood	
Evacuate	0.3(+2) = 0.6	0.7(+2) = 1.4	2.0
Stay	0.3(0) = 0	0.7(+4) = 2.8	2.8

C. Subjective probability

Action	States of nature		Expected utility
	Flood	No flood	
Evacuate	0.8(+2) = 1.6	0.2(+2) = 0.4	2.0
Stay	0.8(0) = 0	0.2(+4) = 0.8	0.8

ties derived by the farmer in the absence of a reliable forecast. Being conservative he has attributed a probability of 0.8 to the occurrence of flood. The outcome now is to evacuate which, although based only on the farmer's own assessment of probabilities, is a rational outcome of his calculations.

Trends in hazardousness

On the basis of a far-reaching survey, Burton, Kates and White (1978) have estimated that 'the natural environment is becoming more hazardous in a number of complex ways that defy immediate or easy reversal of the process'. In making this assertion they are careful to make a distinction between loss of property (which has continually risen) and loss of life (which has been reduced as a whole though the effect of individual catastrophes has increased). They also make the point that it is not the global number and occurrence of extreme natural events that is to blame: these remain quite stable, with the odd random fluctuation in the annual rate of occurrences. Rather it is the combination of changing human use systems in conjunction with these natural events that has resulted in disasters of greater magnitude. Society has tended to put itself (or some of its members) at greater risk. However, before examining how this has occurred – and why it is likely to continue – the point is made by Burton *et al.* that the property losses due to exposure to hazard have a complementary side. That is, they tend to be incurred in pursuit of economic gain. In a 'cost–benefit' analysis it may be that the net addition to global welfare is enhanced despite the occasional disaster, though that is small consolation to sufferers.

Burton *et al.* find two broad reasons for this trend towards increasing vulnerability. Increasing population and the spatial distribution of population provide the first. Growing concentrations of people are locating in hazardous areas. Perhaps the clearest lesson of all is provided by the city of Managua, devastated three times in a hundred years by earthquakes but each time rebuilt on the same site, and continually increasing its population. At the time of the 1931 earthquake it had about 45,000 inhabitants; in 1972, when it was struck again, it had 405,000. Secondly, the process of economic development, especially in the transitional period between traditional activities and land-uses and modern urban systems, appears to be accompanied by increased hazard. Often rapid economic growth outstrips the provision of infrastructure and organisation which could alleviate the impact of a natural event whether it be a sudden one like a flood or earthquake, or a gradual one like a drought.

Building upon their earlier work, Burton *et al.* devote much of their attention to the range of responses to hazards. They do this in various ways. One is through the spectrum of geographic scales: individual, community, national and international. Another is through types of response (absorption, acceptance and reduction of loss and economic change) as illustrated in Fig. 6.4. Their contention is that people engage in behaviour that combines adaptation to extreme events with both incidental and purposeful adjustment. However, adaptation and the different forms of adjustment come into effect at different thresholds or levels of stress. At the *awareness* threshold a society arranges its means of accommodating losses, either by leaving it to individuals to bear them or by creating institutions for sharing them (e.g. insurance, disaster relief). The *action* threshold involves a more positive response to

reduce loss through the modification of events (which so far has not been universally successful and also creates side-effects) or by preventing injurious effects. The latter may involve early-warning systems, evacuation, control works, building designs, cropping practices, and so on. Lastly, the *intolerance* threshold prescribes radical actions in the form of changes in land use or relocation away from the hazard. However, human beings seem more able to put up with hardship than with disappointment and it is not uncommon to find people clinging to their land even under the most hazardous conditions and, at the same time, bringing pressure to bear for other forms of adjustment.

Lastly, in the face of what appears to be growing hazardousness, consideration is given to policies. Burton *et al.* present a balanced assessment of three attitudes towards the causes and hence, by implication, the mitigation of disasters. These are: (a) that disasters are purely due to natural events which can best be dealt with by concentrating emphasis on natural-science research; (b) that they are aggravated (in magnitude at least) rather than ameliorated by technological change and, hence, a more sensitive use of technology is required; and (c) that they are socially induced through economic inequality (the poor suffer most), and the under-evaluation of risk. Their view is that none of these approaches is adequate by itself, that each contributes part of the answer, that all three are interrelated, and that there remains much to be learned about them.

Economic uncertainty and risk

Even if we knew all there was to know about the physical environment we should still find it difficult to escape from uncertainty. All choices and decisions contain some element of expectation of benefits to be yielded in the future. Prospective benefits can only be certain if the environment is wholly controlled by the decision maker or if the act and its consequences are simple and internalised in the sense that it is not significantly dependent upon the actions and responses of others. While this is usually achieved or approximated within organisations, there are often instances – particularly where innovation is involved – in which dependence on an imperfectly known economic environment has a crucial influence on choice and decisions. In the case of an investment this may take the form of over-insurance. The future may be more heavily discounted than conditions actually warrant and the investor will seek to 'break even' over too short a period. Or, the opposite may occur, with the investor failing to take account of hidden costs which may lead to failure.

Economic uncertainty and risk occur for several reasons, all of which are interrelated. The most obvious and familiar are temporal variations in costs and opportunities. Trends and cycles contain an element of unpredictability which inevitably gives rise to a range of expectations. (In turn, the parameters of this range influence future trends and cycles.) All economic indicators show some temporal variation and, to complicate matters, many of them exhibit 'change within change'. That is, short-term fluctuations are nested within medium-term cycles which, in turn, are contained within longer-term trends (Sant 1973).

The consequences for economic management – by households, firms and governments – are paradoxical. The most efficient use of their resources requires that they anticipate the magnitude, timing and duration of change and synchronise their actions with appropriate phases of change. However, it is often only possible to identify these parameters when they are actually happening: hence much economic management takes the form of responses rather than pre-emptive action (Prest 1960). The second element of paradox is that private and public objectives in the face of economic fluctuations are likely to diverge. For example, in a phase of demand inflation, private interests are served by buying goods (and thereby fuelling inflation) while public (Treasury) policy is usually to contain or eliminate inflation by reducing demand through monetary or fiscal measures.

A second source of uncertainty lies in the nature of competition. In a world where resources are scarce, firms and other decision-making bodies have to formulate their strategies knowing that whatever objectives they pursue they are likely to conflict with the objectives of others and, possibly, to evoke countervailing strategies. However, countervailing responses are not always predictable: an important aspect of competitive strategies is to keep one's opponent in a state of uncertainty so that his actions are constrained by his own ignorance. Often, in the past, the possibility of damaging competition has been combated by protective devices and co-operative strategies, in the form of cartels, base-point pricing and resale price maintenance. Some, like the international commodity agreements used by de-

veloping countries to stabilise their export earnings, have been considered desirable. Others, used to protect the market share of producers within a country, have earned opprobrium and been outlawed because the cost of their protection has to be borne by consumers.

Thirdly, spatial variation in costs and opportunities, coupled with spatially limited knowledge, contributes to uncertainty on the part of consumers and producers. The state of knowledge about locational advantages and disadvantages has been the subject of a number of empirical studies. An example, dealing with uncertainty (or ignorance) in a situation where conditions are clearly defined, has been documented by Green (1977). His survey concerned the knowledge that industrialists had of which parts of Britain were designated as assisted areas and in which, therefore, there were subsidies for investment. Although the process of learning is not elucidated, it is clear that firms which had been through the act of relocation and had experience of regional policy had a more accurate view of the map of assisted areas – especially of the area in which they were located (Fig. 6.7). It could be asserted that this aspect of uncertainty should never arise. The information about areas designated for assistance is there for all to see. Maps are published in newspapers and business journals, and government departments produce and freely distribute 'information packs'. The lesson might be that the transmission of information needs to be improved: indeed this is a crucial aspect of policy implementation where a government needs to divert private decisions in a particular direction. Yet, at the same time, it would be unrealistic to expect all firms to have complete knowledge of all the variations that might conceivably be important to them one day. When actions are unique, 'once-off' events, it is likely to be more efficient to collect information as and when it is required; that is, when the pressure to act occurs. The task of gathering information may yield extra information, but it also involves additional costs. If firms aim only to satisfy certain objectives rather than to maximise profits (Townroe 1971), then only a *limited* reduction of uncertainty is to be expected. The same is true of other forms of locational behaviour among households as well as firms and public agencies.

Returning to temporal variations, we may begin by noting the relationship between the failure of enterprises and the changing nature of economic conditions. While failure may not actually be a measure of uncertainty or risk, it does indicate that the outcome

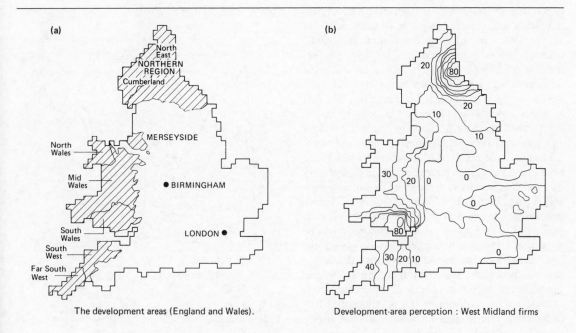

(a) The development areas (England and Wales).

(b) Development-area perception : West Midland firms

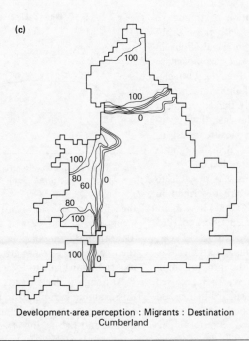

(c) Development-area perception : Migrants : Destination
Cumberland

Fig. 6.7 Industrialists' perceptions of regional policy in England and Wales (Green 1977)

of investments or plans is far from guaranteed. 'Failure' takes many forms and affects many kinds of activity. We can think of the corner shop going bankrupt due to competition from the district supermarket; or the regional plan failing in the sense that its forecasts do not match up to the eventual realities, resulting in misallocation of resources. The first is readily identifiable; the second is likely to be difficult to disentangle. A study by the author (Sant 1975) focused on a select group of industrial establishments (those involved in relocation or which had been opened as branch plants) in Britain during 1945 –71 and comprised a systematic analysis of aggregate and cohort closure rates. The aims were to identify patterns of closure and to relate these to regional policy. It was recognised that some closures might be a mark of success: a firm might have grown to the point where its space demands were such that a new site was necessary. However, the main factor was considered to be failure to achieve satisfactory levels of production and profit at the new location. Evidence exists that it takes up to several years after a move for a firm to achieve either a profitable existence or one that is as efficient as the operation at the previous location (Luttrell 1962). These can be considered as cost factors. In addition it might be thought that recently moved firms with those problems would be vulnerable to temporal variations in demand. A cyclical downturn in sales would cause cash-flow problems which might then lead to closure. Finally, it was thought that failure rates would be affected by external policies. If the government acted to subsidise costs or ease the cash-flow problem then this should show up in closure rates. In fact, these actions formed part of British regional policy, with increased assistance after 1963.

With annual and regional data it was possible to test these ideas and the patterns are described in Fig. 6.8. First it can be seen (Fig. 6.8(a)) how many firms moved between regions between 1945 and 1965 and how many subsequently closed in each year. Since this picture of closure is cumulative the rate of closure is also expressed (b) as the proportion of 'movers' existing at the start of a year being closed during the course of that year. A small but marked fluctuation in closure rates corresponds roughly with the general economic cycle of production and employment in the British economy. Thirdly, analysis of cohort closure rates (c) shows that vulnerability is greatest in the fifth year but stays at a high level until the eighth year before it begins to fall significantly. Lastly, the effects of regional policy can be seen in

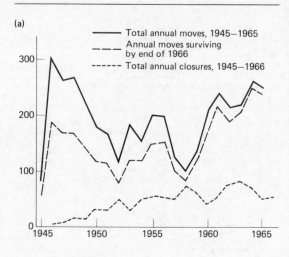

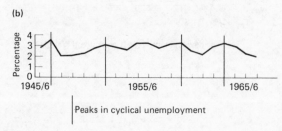

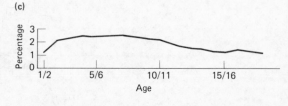

Fig. 6.8 Inter-regional industrial movements and the closure of mobile firms (Sant, 1975)

the closure rates for firms moving to assisted area locations and non-assisted area locations (Table 6.3). The strengthening of policy after 1963 not only made the assisted areas more attractive but also made them less risky for firms to locate in.

Although it deals with a specific set of conditions, this example permits us to make an important generalisation. As in the discussion of environmental hazard and risk-taking strategies above, we can attribute failure (which is analogous with natural disaster) to the combination of external, temporal variability and internal assessment of risk associated with the vulnerability of the enterprise.

Studies of economic variability, with emphasis on

Table 6.3 Adjusted closure indices 1945–71 in peripheral and non-peripheral areas (UK = 100)

Areas	1945–60	1961–65	1966–71
Peripheral*	117	100	91
Non-peripheral	83	100	107

* Peripheral areas are defined as Northern Ireland, Scotland, Wales, Northern England, the North West and South West standard regions.
(DTI unpublished statistics)

business cycles, have a long history. Seen in historical perspective, their content has generally been related to the problems of the time and place at which they were carried out. In the 180 years since Herschell (1801) postulated a causal chain from solar variations ('sunspots'), through harvests and food prices, to demand for non-agricultural goods, there have been put forward at least five cyclical patterns and causes (Sant, 1973). These include innovation waves (50 year periodicity), construction and population (20 years), agricultural (10–15 years), investment (7–11 years) and inventories (3–5 years). Not all have been accorded equal credence. Also, while there is little doubt that cyclical forces continue to dominate economic development (Bronfenbrenner 1969), when economic time series are disaggregated by sector (Mitchell 1951; Mitchell and Deane 1962) and by region (Sant 1973; Haggett 1971), it is clear that no two cycles are identical in their parameters. Amplitudes and duration have proved variable, as have sectoral and regional impacts.

In addition to changes that are strictly cyclical (i.e. having a built-in tendency for reversal), there are changes that are seasonal and structural. The former can be predicted within a broad range, though they are still subject to extreme events as discussed above. Structural change involves a permanent deviation in trend or a sharp shift to a different set of operating or market conditions. Such changes have a variety of forms, reasons and consequences. Major examples include exhaustion and discovery of resources, changes in policy, technological innovation and changes in lifestyles and preferences. Lastly, there are 'random' events which have no direct lasting effect (though they may induce structural changes) but which may interrupt operations sufficiently to affect the costs and benefits of a project.

Issues of forecasting are discussed in Chapter 7; at this point the aim is to focus upon means of coping with economic uncertainty in the planning of projects. These can be divided into two groups: 'rule-of-thumb' methods, and probabilistic assessments. The crudest way of dealing with uncertainty is to set a cut-off period within which to recoup costs so that even if failure occurs later at least the project will have shown a net benefit. In very risky projects (e.g. the production of fashion items) one finds the cut-off being reduced to little more than a season, and goods being priced accordingly. In contrast, some public projects (which are unlikely to be jeopardised by cash-flow problems or competition) can be given appreciably longer cut-off periods up to several decades. However, even this may be insufficient to 'break even' and it may be necessary to allow for the possibility of economic hazard within the cut-off period. This can be done either by adding a premium to the rate at which future benefits are discounted or by arbitrarily revising the expected prices of inputs and outputs to the project.

Probabilistic assessment of risk requires a statement of the likelihood of different events and of the outcome of alternative choices. Conceptually, this approach can be divided into *objective* and *subjective* probability. The objective approach is concerned with repetitive events. As with the tossing of a coin we cannot say with certainty which event will occur next, but we know the probability of each event occurring. Then, assuming that we can predict the outcome of each action for every 'state of nature', the probability of each 'state' can be used to predict the probability distribution of outcomes. Many decisions do involve events and conditions that are sufficiently repetitive – and about which there are sufficient data – for objective probability assessment to be a realistic approach. Most decisions concerning seasonal activities fall into this category. Many consumption decisions also have this rhythmic pattern, though they often possess a range of seasonal fluctuation.

In other cases there may not be a frequency distribution of past experiences, either because there is no record or because the event is not repetitive. When this occurs there is a need to rely on subjective probability – which, in reality, may be no more than informed guesswork. Effectively, this approach requires that *belief* in the possibility of an event be identified. An analogy exists with the notion of 'willingness to pay', discussed earlier (p. 24) in which the interpretation of belief is based on the willingness of a person to act upon the belief. Simplistically, this can be restated by saying that if a person *believes* a risk to be small then it *is* small as far as he is concerned, and vice versa. Clearly this is not very satisfactory, since what a person believes, and is prepared to act upon, can be conditioned in many ways. On the other hand, if actions *are* based on subjective probabilities then, for all its paradoxes, the concept is very valuable. Firstly, since evaluation of risk nearly always (in reality) incorporates a subjective element, it is important to make this explicit. Secondly, it is possible to compare subjective probabilities attributed by different people to the same problem. Some are more risk-averse than others: they may have a moral distaste of gambling, or they may lack the resources to sustain a loss, and so on. A hypothetical example is described above in Table 6.2.

The application of concepts of uncertainty and risk-taking in geography has already been discussed in the context of environmental hazard. Here the emphasis lay on the choice of land-use strategies where resources were locationally fixed. Another aspect of the concept's application is in the location of activities (especially footloose ones) and the way in which they operate once they have been located. This has been subjected to a lengthy theoretical investigation by Webber (1972). It is not possible to summarise all of Webber's arguments and conclusions but his major ones have considerable significance for development and planning. His central theme is that uncertainty leads to conservatism and that this, in turn, affects the economic landscape. Firstly, it reinforces agglomerative forces, with firms preferring central locations over and above what is justified by locational costs alone. At the same time this reduces the number of firms in an industry because a central market can only support a given number; and distance from peripheral markets reduces sales to them. Secondly, the correspondingly greater risks to firms that do locate peripherally offsets the advantages of having little local competition, and they tend to underinvest. Reinforcing these, the third point is that uncertainty affects distance costs, making it more expensive to operate in the already more uncertain peripheral locations: for example, firms have poorer communication linkages and need to carry larger stocks.

Webber's theoretical arguments appear to have empirical support in the way that firms' locations and spatial relationships evolve through product cycles (Vernon 1966; Hamilton 1974). They also have planning implications in so far as this can help to reduce uncertainty by the creation and improvement of information services. Of the three major sources of uncertainty (the behaviour of competitors, the state of the environment, and technical innovation), the last two are areas in which public involvement is normally applicable and socially acceptable.

Summary

Despite all the uncertainty and risk that occur within our natural and social environments, life still has to go on. People have to make a living, often in the face of hazards, and investments need to be made which may have working lives far beyond the foreseeable future. An economist would make the further point that, in terms of total welfare, the failure of a firm or the occurrence of a natural disaster are risks that when spread across a whole society usually have a relatively small impact. If, after investing $1,000m. in a project in the USA, the whole thing is called off, the loss only represents $5 per capita. Moreover, other projects may yield unanticipated benefits due to some unforeseen set of circumstances.

But while risks may be small compared with national systems, the same is less true of their distributional effects. National hazards and economic vicissitudes are often concentrated, spatially and socially, in their impacts. Consequently, 'living with risk' requires at least two lines of action. One is the ability to compensate sufferers when things go wrong. The other is a longer-term response – to reduce vulnerability by careful land-use and locational planning. Both of them have to be backed up by effective information systems, the subject of the next chapter.

7 Information, indicators and forecasts

There is some ambivalence in academic attitudes to information systems, for the excitement they evoke lies in the pursuit rather than in the quarry. An information system is a formal, purposive arrangement for the collection, analysis, display and collection of facts, but yet is substantially less than original research. Innovation comes in the design of the system, rather than its use, and it is in the former that the greater intellectual challenge arises. Once established, an information system exists as a way of assembling and interpreting information according to some pre-ordained scheme. At this point the application becomes repetitive and, for the innovative researcher, much of the interest lapses, save perhaps for attempts to modify and improve existing systems.

However, it should not be concluded from these comments either that there is some fixed set of information systems just waiting to be defined, or even that much progress has been made towards establishing fully-fledged, workable systems. Rather, what we have (and need) is a substantial body of research and experiment at scales ranging from the global to the local, dealing with social, economic and environmental conditions with the objectives of reducing uncertainties and assisting in resource-allocation decisions.

In this chapter the initial emphasis is on the structure and content of information systems, with some discussion of types of system and of the 'input–transformation–output' format which they conventionally employ. Attention is then shifted to two of the major purposes of information systems, namely monitoring and forecasting.

Information and communication

For most people, most of the time, information poses few problems. They can communicate quite adequately in an informal way with confidence that their messages will be understood and evoke a response. Moreover, their informal systems can happily contain subtle nuances, even ambiguities, which will have meaning to a listener or reader. Informal systems are very powerful elements of culture.

However, there are times, particularly in original research and in planning, when an informal system has to be replaced by a formal one. These occur when the ordinary rules of a language are augmented by other criteria, dealing not only with *what* is communicated, but also with *how* it is communicated. To some extent the rise of formal systems has been the outcome of computer technology, but it would be incorrect to think of them purely in terms of programming languages. Indeed, very little of what follows relates directly to this aspect of information systems.

The study of information, or semiotics, has a number of interrelated aspects that concern applied geographers. Each of these aspects deals with different ways of understanding signs, symbols or signals and of measuring their properties. They include (following Stamper 1973) the following: pragmatics, semantics, syntactics, and empirics.

Pragmatics deals with the relationships between information and behaviour, and examines the social contexts of the interpretation and response to information. The situations (though perhaps not the usage) where pragmatics have been examined by geographers are familiar. For example, in the diffusion and adoption of innovations (Blaikie 1978), the major pragmatic questions included the sensitivity of people to information, the means by which it is exchanged, and the role of particular individuals in the transmission of information.

Semantics is concerned with the social meaning of signals, their cognition and interpretation by receivers (e.g. in disentangling facts from value judgements), and their manipulation by producers of information to convey particular types of message. Choices between statistical techniques to use in analyses, or between verbal and visual presentations, or between different ways of phrasing a questionnaire all fall within the purview of semantics.

Syntactics takes us into the heart of formal systems dealing with the logic and organisation of storing, retrieving and combining signals. In this the role of computer and meta-languages are important, but geographers have been dealing with syntactic problems long before the advent of Fortran. Every map – thematic and topographic – poses syntactic problems. New syntactic problems are created by the increasing use of modern remote-sensing techniques such as false-colour air photography and satellite imagery (Cooke and Harris 1970; Barrett and Curtis 1976). The output from these devices is usually a representation of environmental phenomena in the form of 'spectral signatures' or images which record on film the intensity of electromagnetic energy transmitted by each object. Their interpretation requires knowledge of the correspondence between spectral signatures and the physical properties of different phenomena, which may vary for a number of reasons. Apart from technical causes (e.g. type of film, wavebands used for recording images, angle of aspect), signatures may also be affected by seasonal variations, types of terrain, and neighbouring features. When all of these have been resolved, remote sensing is a very valuable tool for primary data collection.

Empirics deals with the more strictly technical problems related to the mechanical and physical properties of signals and their transmission. Though interesting, it is not an aspect of information systems that is directly the concern of applied geographers.

All of these branches of information science are relevant to the design of information systems but, as we might expect, with varying degrees of importance. Thus in collecting and storing data the syntactic properties might be emphasised, while in presenting output to decision makers the semantic properties might take precedence. In a third case – attempting to disseminate information – greatest attention might be paid to the pragmatic aspects.

Cutting across these branches, Stamper adds a typology of information that also concerns us. Information can be used with different intentions and in different modes. These were summarised in Table 1.1. Two familiar intentions are noted (description and prescription) and two modes (denotive and affec-

tive). The classification is a useful one, because it can be related to the ways in which geographers (as pedlars of information) have worked in the past, and work now. Apart from the fact that geography has never been very strong on affective-prescriptive information, the scope of present interests in the discipline clearly embraces all of the other three classes. But the table has a further significance. Each mode and intention requires a different set of semantics and also a different approach to the (pragmatic) question of presenting information to social groups. For example, the reader might consider the methods necessary to persuade a person or group of people to accept: (a) facts, and (b) value judgements. It would be surprising if the answer was the same for both types of information.

Information systems

Information systems are a special form of communication system. In more common usage they consist of a structured flow of 'signals' (which may be either quantitative or qualitative information) directed towards a particular user, for whom the information is a necessary input for making decisions or conferring advice or other information. The attributes of 'structure' and 'direction' are crucial for two reasons: first, they impose a limitation on what is to be contained in a system; and, second, they are themselves defined in practice by the needs of the user. Hence, while the principles may be similar, the form and content of each system will depend on the purpose to which it will be put.

Simplistically, an information system can be represented in the form of a continuous flow diagram (though the continuity may be more apparent than real) in which inputs (primary data) are transformed by appropriate analysis into outputs which then may provide a basis for action (Fig. 7.1). Alternatively, the output may be returned to become a new input for further analysis. An example of the latter occurs in many monitoring systems where conditions found at time t_1 influence conditions at t_2. In general, therefore, an information system is a funnelling and filtering process whereby large amounts of data are reduced to prescriptive or descriptive statements. However, this simplistic view masks many of the conceptual and practical questions related to the establishment of information systems.

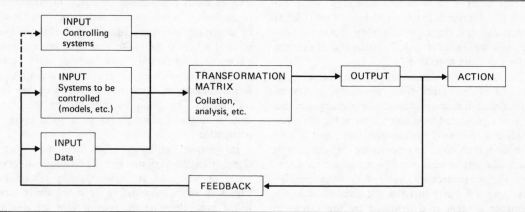

Fig. 7.1 Elements in a formal information system

Though it may appear to be illogical, the most appropriate starting point in designing an information system is not with the inputs but with the needs of the user. Those requirements set the conditions for defining the rest of the system. As Berry (1972) has effectively, if inelegantly phrased it, the relevant questions are: 'What do we know? What do we need to know? and how do we find out?' For the user, the desired information system may take one of several forms. Sutton (1977) has defined three basic types ('hot check', management and investigative systems) to which we can add a fourth, which is a hybrid of these three.

Hot-check systems aim to systematise the response to unique events. Sutton, who was concerned with criminal-investigation procedures, used the example of automobile theft (every car is unique even if theft is fairly common), but there are also cases in environmental science (e.g. natural hazards) and in urban planning (e.g. building and development applications) which, in theory at least, need to be met by a rapid response.

Management systems operate like a continuous inventory in a warehouse, whereby changes in outputs automatically, or semi-automatically, affect the rate of input and vice versa. Again there are many examples in environmental and economic planning: the flow of water through a dam in relation to precipitation, or the provision of school places in relation to the number and distribution of children.

Thirdly, investigative systems are concerned with basic, innovative research into processes and patterns which, in turn, provides the intelligence on which policies are made. This includes – though the point is easily overlooked – the research which may be necessary to establish efficient hot-check and management systems.

The fourth, hybrid, type can be called the planning information system. Here the needs of a user tend to be quite diverse, particularly as one moves further into indicative planning (see p. 12). Development control, budgeting and resource allocation, zoning, provision of infrastructure and services, long-term forecasting, and so on: each of these functions imposes a different demand on information systems. In theory it might be possible to incorporate them all in an integrated system and some success has been achieved in this direction in urban-planning systems (see p. 90) but in reality it is questionable whether total integration is either feasible or desirable. Information systems are expensive to establish and, once established, expensive to modify. So, it may be that flexibility is a more desirable attribute than exhaustiveness.

There arises at this point a crucial conceptual issue concerning the parallelism between information systems and 'real-world systems'. The problems here are the extent to which human and environmental processes actually are systemic within the terms of general systems theory (i.e. working together through a regular set of relationships) and, if they are, the form that the systems take. The extreme argument is that all objects are systemic, that there is a hierarchy of complexity from closed, static inorganic systems to open, dynamic, organic ones, and that if we search long enough then we ought to be able to discover how each system works (Kast and Rosenzweig 1972). Conversely, it has been asserted that the concept itself is ill-founded, especially when extended to human activities (Chisholm 1967). What undermines it is the unpredictability of internal relationships within the so-called system and of external influences from outside it.

The relevance of this debate is considerable. If, in fact, no system exists then, obviously, it has no equivalent information system: the phenomenon can only be represented by disjointed facts. On the other hand, if a real-world system does exist, and can be identified, then it can be represented (subject only to the availability of resources) by an information system. For the present we shall avoid these issues. Instead, we will assert that the effectiveness of an information system is determined by the extent to which a real-world system has been established. If this means that many areas of environmental and human activity are not and, indeed, may never be amenable to information systems then this is something that has to be recognised. It does not negate the value of research to establish the dimensions and processes of a system, but it does protect against the setting up of spurious information systems.

To illustrate these points we can begin with the hypothetical and simplified example of the management of a warehouse. Here the real-world system is the flow of goods: the relationship between inputs and outputs is readily established (if we ignore the vicissitudes of economic booms and busts). Every flow of goods is paralleled by a flow of information in the form of written orders and despatch notes – or by electronic messages. Indeed, the point has been reached where some major retailers have automatic electronic information flows between sales outlets and warehouses (Berry 1980) thus integrating information and real-world systems to almost their fullest extent. The revolution wrought by the silicon chip is doing similar things to a wide range of occupations.

In contrast, while we might have more than a hunch that the various aspects of regional development or land use are systematically related (see Fig. 5.1), often it is still to be conclusively established precisely what the relationships are. For example, in Table 7.1 we present an abstract of statistical results from studies of various aspects of employment. It is questionable whether any of them forms the basis for forecasting the distribution of change, let alone its magnitude or timing. Perhaps we should expect no more, since all of them are partial analyses and perhaps further research will lead to much better explanations. Neither does the lack of established real-world systems preclude experimentation, using a systems framework to order the information used in research. The heuristic value of systems theory should not be overlooked even if its applications are limited.

Although it is natural to concentrate upon the

Table 7.1 Summary of statistical results from a sample of studies of selected aspects of employment (Sant 1978)

Author and year	Region	Area units	Variables	Period	Correlation
Sant (1967)	GB	74 cities'	Unemployment (%) v. industrial structure	1960–63	0.23 to 0.49
Smith (1968)	NW Eng.	95 areas	Unemployment (%) v.	1966	
			(a) change in total employment		0.24
			(b) no. of industrial employees		−0.59
Tulpule (1969)	Greater London		Dispersion of industrial employment v.	1951–61	
			(a) growth of net output		0.55
			(b) investment per capita		0.34
			(c) female participation		−0.34
Hart (1970)	E & W	10 regions	Gross migration v.	1960–61	
			(a) 'gravity model'		0.66
			(b) 'gravity model' plus rate of industrial building		0.75
Gordon (1970)	GB	62 sub-regions	Female activity rates (age adjusted) v.	1966	
			(a) unemployment (%)		−0.62
			(b) average earnings of male manual workers		0.23
Keeble and Hauser (1972)	SE Eng.	112 areas	Change in mfg employment (%) v. male unemployment (%)	1960–66	0.42
			Change in mfg employment (total) v. change in population (total)	1960–66	0.57
Gordon and Whittaker (1972)	E & W	55 counties	Net male migration, aged 15–44 v.	1961–66	
			(a) average male earnings		0.40
			(b) unemployment (%)		−0.34

operational aspects of information systems there are, as Hermansen (1968) has argued, two other sub-systems to be considered. These are referred to as: (a) the system to be controlled; and (b) the controlling system, and are discussed in terms of regional economic planning. Hermansen defines the 'system to be controlled' as the set of social and economic processes whose interrelationships bring about the distribution of developments, and the 'controlling system' as the administrative and executive framework responsible for planning and control.

The separation of these two sub-systems appears to work better in concept than in practice. Ideally, in representing relationships and processes we would concentrate upon their intrinsic attributes regardless of their spatial dimensions. If processes operate locally, or involve individual households or firms, then this is the appropriate scale of analysis. Alternatively, a macro-economic analysis might be required, using data at a regional scale. Or it might be that a mixture of scales is needed. However, in reality, the dimensions of the controlling system also often control the form in which data is collected and made available.

Two features tend to dominate the character of information. Firstly, each agency collects what it considers to be relevant, in a way that is appropriate to itself. Secondly, each level in a hierarchical system tends to concentrate on information about itself and about the next level down: central government on local government, and local government on land parcels. Putting these two generalisations together, the usual outcome is that data sets tend to be internally consistent (subject to changes over time in the definitions of areas or objects) but that the data collected by different agencies are often incompatible with each other. Even where this is not so, it may still be necessary to adjust data sets before they can be used in the same analysis. Commonly this involves a process of aggregation in one data set to bring it into line with the other, which means that detail has to be foregone.

The determination of these two sub-systems is a necessary condition for the establishment of an effective information system, but they are no more than a beginning. It is still necessary to have a clear understanding of the nature and objectives of the inputs, transformations and outputs of the system.

Inputs

It has often been pointed out that the last two or three decades have witnessed an 'information explo-sion'. Censuses have been taken in countries where previously there were only unreliable estimates. (Sometimes these censuses are a bit suspect, too – Mabogunje, 1976.) Countries which already had regular censuses now have them more frequently and ask more questions. In Australia, for example, the 1976 census asked 41 questions about personal attributes and 12 questions about dwellings: in 1954 the respective totals were 14 and 8. The same trend has been repeated in other countries and largely reflects the response of census bureaux to the requests of census users. At the same time it has provided a 'backlash' from those who have to supply the answers: complaints about invasion of privacy, and poor response rates even to official surveys, have become commonplace at census time. In addition, many countries now have a large number and variety of specialised surveys and monitoring systems dealing with economic and social conditions, many of which are readily available in statistical abstracts. Thirdly, remote sensing has provided the means for data collection on an even more massive scale (Allan 1978). LANDSAT imagery, with its repetitive satellite coverage of the earth's surface (albeit at a rather coarse level for many researchers), has supplemented our ability to map and interpret land use and land coverage, particularly in more remote regions (Rudman 1977). Other instruments for remote sensing monitor and measure conditions as disparate as atmospheric pollution and vehicular traffic, seismicity and water supply. Fourthly, and not to be overlooked, is the plethora of documentary information – ranging through pieces of legislation, government papers, private reports and academic publications – all of which may be relevant to an understanding of controlling systems or systems to be controlled even if they are not direct inputs to an information system.

Many of these data have quite specific uses and users, and their attributes (periodicity, sample size, geographical coverage, etc.) deliberately meet these needs. Yet there sometimes remains a hiatus between what is available and what is desirable. This often occurs after the passage of new legislation, the implementation of which may be impeded unless information systems are quickly contrived. For example, environmental legislation requiring the preparation of environmental impact studies (such as the US National Environment Protection Act of 1969 or the New South Wales Environment and Planning Act of 1979) has created a new demand for information systems which, in turn, has been met not only by a wide variety of proposed techniques but also by different views of the form that information inputs should take. Thus a contrast may be drawn between Leopold's matrix technique which uses ordinal (scaled) measurements of the magnitude and importance of impacts, and the Battelle technique which aims to use cardinal measures (Canter 1977). These are discussed further in the next chapter.

A hiatus also tends to occur in the making and monitoring of plans and policies, especially at the regional level. 'Regions' and 'spatial systems' rarely have a statutory basis in a comprehensive sense, though they may have a special-purpose role. In consequence they are almost certain to lack the machinery for comprehensive data collection. Thus when a regional planning team is set up it usually has to assemble a set of regional statistics for a one-off job. The difficulties soon become apparent. Statistics, if they are available at all, are available on different areal bases; with different degrees of up-to-dateness; with different periodicities; and with variable accuracy. Frequently, also, the recommendation of planners, that there should be continuous monitoring of trends in their regions, is met only by a centralised collection of existing statistics without relating them to the aims and forecasts of plans.

Transformation

Without doubt the transformation matrix is the engine-room of an information system. It is here that the inputs, which mean little by themselves, are interrelated with each other in order to distil the essential picture that is necessary for decision making. It is the analytical stage of an inquiry. However, in a practical sense it is more than the pursuit of a general view of the patterns and processes involved. In Hermansen's (op. cit.) terminology, the analysis forms the meat in the sandwich between the controlling system on one side, and the system to be controlled on the other. This imposes a discipline that is rarely found in pure research. (Note that such discipline is not virtuous *per se*: it is merely necessary in this context.) Thus the inputs to the transformation matrix include: (a) *'system information'*, related to the administrative or executive functions of the region or organisation of the controlling system; (b) *'operative information'*, concerned with the models and methods that may be of use in analysing the patterns and processes of the system to be controlled; and (c) *'data'*, the variables and attributes to be analysed.

Assuming the system information and the data are readily accessible, the key to a successful transformation matrix obviously lies with the operative information and the skill with which it is applied. In inventory systems, where the interrelationships are well established, the responses can be almost as automatic as a plumbing system. Information goes in about the rate of use of a stock, is related to data about remaining stocks and likely future use, and an inference drawn about the need to regulate future use or to replenish stocks. Public utilities provide examples of these; so, though perhaps to a lesser degree, does traffic management.

The same cannot be said for investigative systems. Although an aim of most research is to derive models which could be the basis of transformation matrices, the stage is still far in the future when social engineering will be subject to the same degree of automatic communication, analysis and response. The reasons are not hard to find.

Firstly, there is the problem of defining precisely the objectives that the analyst is trying to serve; different objectives not only lead to different data requirements but also to different methods of analysis. To illustrate this, consider the choice of technique for evaluating a public project – say a new routeway. If the objective is to maximise internal benefits to the operator then a cost-effectiveness technique might be applied, whereas if the objective is to affect the distribution of welfare a more appropriate approach might be through cost–benefit analysis.

Secondly, questions arise about the applicability of models derived at one time and place (or in one spatial system) to situations that are dynamic or heterogeneous. It has been found, for example, that exponential values in gravity models may not be transferrable from one system to another (Olsson, 1965; 1978) and that input–output models of econometric forecasting suffer from unstable coefficients as inter-industrial linkages change (Allen, 1970).

Thirdly, the effectiveness of the transformation matrix may be reduced because system information is not available or because data is sub-optimal. Examples of weak system information occur in some regionalisation studies. One, in Australia, asked the research team to define a set of regions, which they did – without knowing what the new distribution of powers and responsibilities would be. The result was an interesting use of classification algorithms but without much practical value (Logan *et al*. 1975).

Fourthly, what actually goes on in the transformation matrix may be a highly complex and fragmented set of operations and, possibly, of operators as well. Whether such fragmentation creates insurmountable problems depends partly on the nature of the real-world system and partly on the degree of control exercised over the information system. Homeostasis – whereby static or dynamic stability is maintained despite external forces – makes it easier to define and use an information system. Reliance on many independent agents, each of whom has a different interest in the success of the information system, makes it more difficult. One may draw a contrast between the apparent efficiency of military and civilian information systems in this regard (and, at the same time, suggest that inefficiency is often worth preserving).

Outputs

Having passed through the transformation stage, where the major issues are syntactic, it would seem realistic to suggest that in the preparation and presentation of outputs the emphasis should shift to semantics and pragmatics. At this point the crucial questions concern the interpretation of whatever analysis has been conducted and the use of that information. The purpose of a system's output is to evoke an active response by its users. These may be organisations working towards private or public ends, or they may be a population at large. In either case the requirements are that the output should have a clear meaning in terms of the relationship between action and outcomes and, possibly also, between these and the objectives of decision makers. For a firm this might mean showing the likelihood of different actions resulting in different profits; for a social welfare department the output might be to describe the distribution of need, or to predict the impact of different strategies.

The extension of information systems beyond the single decision-making organisation presents a challenging and sensitive area for applied geography – as it does for all information services. Here the output of a single system (e.g. a weather bureau) may simultaneously become the input of many individuals (e.g. farmers, or holiday-makers, or builders). Or it may be that an organisation is concerned with the efficient but sequential diffusion of information: for example, an agricultural extension service setting up a model farm, which is observed and (hopefully) copied by some farmers who, in turn, are observed and copied by others, and so on.

These examples raise two major questions concerning, firstly, the nature of information and, secondly, the nature of the communication process. Much attention has been given to the latter, through

the study of the diffusion of innovations, though Blaikie (1978) has criticised this work in general as tending to give insufficient attention to the societal environment in which diffusion occurs. His argument is that processes of diffusion reflect more than the access of individuals to information or the spatial structure of channels of communication, and that to understand spatial patterns one first needs to understand political economy. When this is achieved one has the context within which to manipulate information and communication. And herein lies the most sensitive issue, since media can be manipulated perniciously as well as beneficially. Control and use of media are important political weapons. They can be employed for such disparate purposes as fostering xenophobia, capturing support for a political party or initiating a run on a commodity, as well as providing information on economic and cultural opportunities or giving early warnings of natural hazards. But whatever their use, the effectiveness of media is affected, to a significant degree, by the interpretation of their messages which, in turn, is influenced by their semantic and pragmatic attributes as well as by the attributes of 'interpreters'. A simple example is found in the language of weather forecasts (Smith 1975). A forecast that 'rain is likely' leaves much to the discretion of the individual: he has to make his own evaluation of how likely, how much rain, and what the effects are of rain upon his planned activities if he goes ahead with them. A more definitive forecast may not be feasible, but we can imagine that if one were given which stated the probability of a certain amount of rainfall it would have a different influence on people and their actions. Of course, there is always the problem that a sequence of incorrect forecasts may lead to a credibility gap, but that is another issue.

At this point it is worth posing the questions of what factors influence whether or not an item of information is collected and a formal, purposive, information system is asembled. At first these might seem to be rather silly questions: they will be done if someone is interested in doing them. However, in his review of the market for policy-indicators, Rose (1972) suggests some important guidelines that relate to the potential support for, and adoption of, indicators and information systems. His argument is summarised in an equation which has a conceptual, rather than practical, value. New information will be used by an organisation if:

$$U_i > C_o + C_c + C_v + C_a - C_{ia}$$

Otherwise it will be ignored. In this equation U_i is the utility of information, which may include subjective assessments; C_o is the cost of obtaining information; C_c the cost of consuming it (i.e. making 'sense' out of the results and reports of the researcher); C_v is the cost of value conflict, which arises when the user has to change his own attitudes and values before accepting the research; C_a is the cost of action; and C_{ia} is the cost of inaction, both of which might be measured in terms of votes lost or gained, if the organisation is a political party, or in real costs if it is the community at large.

Despite containing some subjective elements this is, nevertheless, a useful basis for evaluating arguments about the virtues of information systems. The first two and last two cost items can be treated objectively in many instances, as also can utility. This would be the case in systems for hazard warning or some aspects of urban management, for example. In these, also, the cost of value conflict would probably not arise. However, where subjectivity does play an important role (e.g. in attitudes towards poverty and wealth) then U_i and C_v are likely to take on a range of values reflecting different political viewpoints, and the implementation – or lack of it – may reflect opportunism (i.e. who is in power) rather than objective benefit.

Example: an urban-planning information system

Many of the points above can be illustrated in the design and use of information systems for urban planning. This has been the subject of a large number of expensive experiments in many different cities around the world. Often systems have failed, or not been effective, not so much for technical reasons as because they paid insufficient attention to administrative frameworks and requirements. The example discussed below, the Sydney City Council Land Information System, has not been free from these problems but has been more successful because its inputs are collected in the normal course of the city's planning and management. By combining the various sets of data in an interactive computer system the result is greater efficiency in day-to-day activities and also a basis for long-term structure planning (Nash 1978).

The structure of the system is described in Fig. 7.2, which shows the sub-systems which deal respectively with finance, planning control, local government services and surveys and mapping. Linking these is a control sub-system which contains the geographical code or property key (PKEY) which locates every parcel of land in the city by its address. In turn this can be related to the content of the manipulation and processing sub-system (MAPS) where a Map Base File contains the digitised boundaries of all land parcels and structures, together with other known areas such as wards, districts, street blocks, postcode areas and census districts. This sub-system also contains survey and census data and is valuable in urban research and mapping. It is one of the two sub-systems that are most interesting in this discussion.

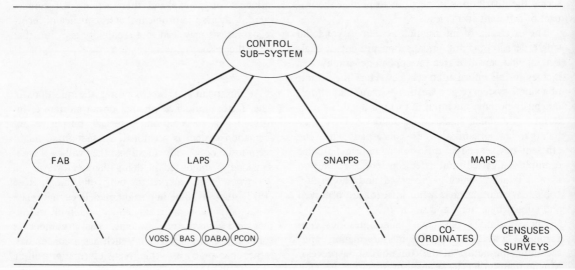

Fig. 7.2 Sydney C.C. Land Information System. FAB: Finance, Accounts, Budget subsystem SNAPPS: Service Applications

The other is the Land and Property Sub-System (LAPS) which contains information on valuation, ownership and sales (VOSS), a history of development and building applications (DABA), planning controls (PCON), and register of voters. Three of the uses of this sub-system can be listed. Inquiries about development and building applications can be answered almost instantaneously and applications themselves can be processed rapidly, allowing them to be handled within the statutory 40-day limit except where major developments and outside agencies are involved. Secondly, information on the value, character and quality of sites and buildings assists in identifying appropriate areas for public investment. Thirdly, the information on economic activities, land use and employment in each building (BAS) provides an important base for urban modelling.

All of this information is subject to continuous updating and can be supplemented by other data as it becomes available. For example, the sample points of a kerbside pollution survey could be fed into the MAPS sub-system for use in conjunction with other data (e.g. traffic counts and land use) already there. However, one critical drawback exists which is common to many similar systems: namely, that they apply only to a limited area. In this case, the City of Sydney is a very small part of a metropolitan region which contains 41 autonomous local government areas. While their day-to-day management and planning may have little relevance to what happens in the City, this is far from true of their long-term development. In an ideal world the City's system would extend to the entire metropolitan region.

Social and economic indicators

Indicators have two main purposes: to tell us where we are, and to point to where we could be going. (Presumably we know whence we have come!) Thus there is a fundamental link between describing the present and predicting the future, in the way that social and economic indicators are defined and used. To Biderman (1966) an indicator is 'a statistic of direct *normative* interest which facilitates... *judgements* about the condition of major aspects of a society. In all cases it is a direct measure of welfare and is subject to the interpretation that if it changes in the right direction, while other things remain equal, things have [become] better or people are better off' (my emphases). This statement raises many

questions: of interest to whom? what aspects are 'major'? should every indicator refer to society in general? what about minority groups? what if 'other things do not remain equal'? how, then, do we define 'better off'? and so on.

Biderman's definition is also notable for its emphasis on real outputs rather than real inputs. It is the condition of society rather than the factors giving rise to that condition that is the basis for an indicator. For example, in judging the educational attainment of a society it is such measures as years of schooling or levels of literacy which are the relevant indicators, rather than educational expenditure. The distinction is an important one for it encourages greater clarity in the use of data. However, at the same time, it is possible for too strict a focus on outputs to have a stultifying effect on the explanation and prediction of processes of change. Also, many indicators have a dual role, as outputs and inputs.

Apart from this distinction between inputs and outputs, indicators have been classified in several ways related to comprehensiveness, use, and ease of interpretation.

In a discussion of social indicators and public policy Knox (1975) asserts that 'in the final analysis... problem-oriented indicators must be seen as stopgap measures. What is needed to assess the effectiveness of public policies is a comprehensive system of indicators which cover all aspects of well-being.' This is a view that appears to have been widely held in recent years, and is evident in the work of a number of adherents of the 'social-indicators movement' (Miles 1975). However, comprehensiveness can mean two quite different things. One is the development of a set of indicators, each of which individually describes a feature of well-being that is considered important. The other is to distil from a number of indicators one unitary measure of (say) the quality of life, with each indicator contributing a weighted share to a single index. The latter approach has attracted a proliferation of multivariate analyses, many of which have little theoretical underpinning and not much practical value except, perhaps, in confirming the importance of individual indicators. We would assert, therefore, that comprehensiveness in the latter sense is not necessarily an alternative to individual, well-specified, problem-oriented indicators.

Turning to their use, Carlisle's (1972) classification of indicators contains the following categories: description, prediction, problem-identification and programme-evaluation. One might expect the demands on data to differ according to these uses. For example, prediction is likely to require substantial time series, whereas problem-identification may be satisfied by data referring only to a single point in time. Programme-evaluation indicators may also need a time series, depending on the nature of the problem, in order to monitor progress towards stated objectives.

An alternative classification, from Kamrany and Christakis (1970), allocates indicators to three categories: absolute, relative and autonomous. Absolute indicators are those for which there is substantial agreement about their significance for human welfare: the notion of minimum acceptable standards for water or atmospheric pollution or housing quality are examples. Relative indicators possess no such 'fixed' conditions but permit comparison between different areas or social classes. Autonomous indicators cut across the other two and are related to specific areas or minority groups.

The interpretation of indicators ought to be straightforward, but difficulties can arise in relating the thing being measured to the conceptual framework of a study. In this respect Cazes (1972) suggests three quantification problems that may influence analysis and interpretation. *Fractional measurements* occur when there is inconsistency between a concept and its operational definition. This may arise because a single indicator is thought to be sufficient: for example, rates of unemployment as a measure of regional welfare. It may also occur in a policy-evaluation study where available indicators refer to inputs rather than outputs: for example, in the evaluation of regional policy it has usually been easier to identify levels of public expenditure and numbers of jobs created than to indicate the beneficial effects on different regions. Secondly, *indirect* measurements refer to data collected for a particular purpose and which can only be inferred to be relevant to some other purpose. Along with this we can include the use of one variable as a surrogate for another when the correlation between the two is suspected to be imperfect. The third problem concerns *collective attributes* in which two pitfalls are encountered. One concerns the treatment of the data-collection units (municipalities, states, etc.) as socially or economically meaningful units whereas most previous experience would suggest that they are heterogeneous. The other is whether the measurement of that unit should be a 'global' one, expressing an abstract figure without regard to how individuals in the unit contribute to the measure (e.g. 'the aver-

Table 7.2 Variables used in the analysis of levels of living in England and Wales, 1971 (Coates, Johnston and Knox 1977)

1 % households overcrowded (more than 1.0 persons per room)
2 % households without exclusive use of all three census amenities (hot water supply, fixed bath/shower, inside flush toilet)
3 % households sharing a dwelling
4 % dwellings with only 1 or 2 rooms
5 % dwellings owner-occupied
6 % dwellings privately rented
7 New dwellings completed per 1,000 households
8 Infant mortality rate
9 Local health services: expenditure per 1,000 resident population
10 Average list size of principal general medical practitioners
11 % students in age groups 15–19
12 Ratio of pupils to teachers in primary schools
13 Professional workers
14 % unemployed
15 Female activity rate
16 % persons aged 0–14
17 % persons of pensionable age
18 % households of only 1 or 2 persons
19 % population change per annum 1961–71 due to migration
20 % poll in local elections
21 % households without a car
22 Index of rateable property values $\left(\dfrac{\text{product of 1d. rate}}{\text{total population}} \right)$
23 Public libraries: expenditure per 1,000 resident population
24 Cinemas per 1,000 resident population
25 Population per social worker
26 Police services: expenditure per 1,000 resident population
27 Divorce rate
28 Illegitimacy rate
29 Child-care referrals per 1,000 population aged 0–17

age man') or whether information about the different members of the unit should form the basis of the measurement.

The use of social and economic indicators is as vital and varied as numerical research itself. Moreover, as public sectors grow in their extent and influence it is inevitable that an increasing demand will occur for systematic surveys and censuses to fulfil the tasks outlined above: prediction, problem-identification and policy-evaluation. At the same time there is a need, as Miles (1975) has argued, to guard against excessive claims by proponents of the 'indicators movement'. For example, one cannot help but feel uneasy about the analysis of levels of living presented in Fig. 7.3, in which 29 variables (listed in Table 7.2) have been distilled, using factor analysis, to provide a single index for local government areas in England and Wales. Leaving aside the methodolo-

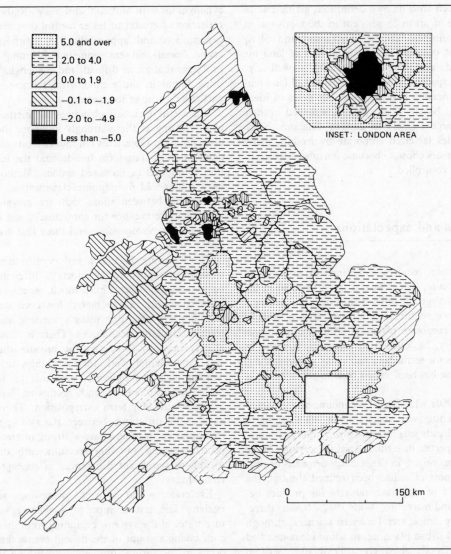

Fig. 7.3 Social well-being in England and Wales in 1971: an index of level of living based on twenty-nine variables. Negative scores are indicative of low levels of living (Coates, Johnston and Knox 1977)

Legend:
- 5.0 and over
- 2.0 to 4.0
- 0.0 to 1.9
- −0.1 to −1.9
- −2.0 to −4.9
- Less than −5.0

INSET: LONDON AREA

0   150 km

gical issues, how are we to interpret these results in terms of their significance for social and economic policy? Would we improve levels of living by changing the proportion of people of pensionable age? or by making voting compulsory? or by increasing expenditure on police forces? or by reducing divorce? Of course, it could be argued that a composite analysis is meant to be no more than a general picture of disparities. But the point remains that a general picture is coloured by what goes into it.

Similarly, it is important to keep a close watch on the syntactic properties of indicators. Much has

already been written on this and most of the points are readily seen. Here these properties mainly concern the periodicity and geographical coverage of surveys, and the definitions used in collecting and reporting data. Attempts to improve the geographical bases of data are found in various national census systems which have introduced grid squares in place of irregular administrative units. However, spatial configurations are less important than the definition of indicators, the outcome of which may be a severe distortion of statistics. For example, in Australia unemployment figures are reported by two federal agencies, each with its own definition, giving rise to a difference of up to 25 per cent in the totals out of work. In another example, Dickinson and Shaw (1977) point out that British definitions of land *use* often include aspects of land *ownership* as well (e.g. *public* open space). Indeed, the questions of land tenure and ownership may also pose problems of identification, as Newby *et al.* (1977) discovered in their work on large farmers in eastern England. Since taxation rules favoured corporate ownership, many of these farmers chose to become tenants of the companies they controlled.

Forecasts and expectations

Rational action requires two kinds of information about the future. One is what is likely to occur in the absence of action; the other is the effect that action is likely to have on future conditions. The growth of public and corporate planning, with their emphasis on rationality, has created an increasing demand in recent years for techniques and skills in forecasting. The response has been varied both in content and in quality.

In the 1960s when social, economic and environmental planning were beginning to gain strength, there was already established a strong body of literature and expertise in technological forecasting. Some of this had its origins in the military demands of wartime, but most of it had been refined through the requirements of corporate industry for product development and marketing (Wills 1972). It was, therefore, a fairly simple step for social sciences, through regional and urban planning, to adopt ideas that had been developed elsewhere. Textbooks written around this time contain several examples of this translation from technological to social systems, often without much inquiry as to the validity of some of the techniques even within technological forecasting (McLoughlin 1969; Chadwick 1971). This is not to denigrate the importance of technological forecasting: it is important in its own right, and the implications of technical change for social, economic and environmental milieux are obviously far-reaching. However, this is very different from assuming that one set of methods and skills can be readily transferred to a different set of problems.

Parenthetically we may note that the conventional distinction between 'hard' and 'soft' sciences is not synonymous with 'difficult' and 'easy'. Rather, it is a reflection of greater or lesser control over events and experiments and applicability of positivistic concepts. Social and environmental planning contain problems that are as difficult and challenging as anything found in the technological sciences. This is particularly true of forecasting.

It is useful to classify objectives and methodologies of forecasting as these strongly influence the choice of techniques. Objectives fall into two categories: the exploration of long-term trends, and the identification of impacts of proposed actions. Methodologies have a four-fold overlapping classification. We can distinguish between those that are *systematic* and those that are *intuitive* (or consensual); and between those that are *extrapolative* and those that are *normative*.

The following discussion will amplify these but it is worth noting where the efforts of different groups have tended to lie. Thus, much recent academic work (not only in geography) has been concerned with impact forecasting using systematic and short-term extrapolative techniques. There is considerable affinity between these and the broader theoretical and experimental work that academics undertake. Among planners, however, emphasis tends to be placed on exploratory and normative forecasting together with long-term extrapolation. There is not an absolute difference between the two approaches (for example, planners have a strong interest in impact forecasting) but they are sufficiently dissimilar to impose significant differences in conception and technique.

Exploratory and impact forecasting are self-evident. The first attempts to elucidate possible future states of the system, beginning from the present and, taking account of trends and events that might occur in the future, ultimately describing the 'scenarios' that might then exist. Its methodology might be any of those mentioned above with the exception of normative forecasting. (A normative approach establishes goals for some future date and then examines the means by which they might be achieved.) The second, impact forecasting, is more circumscribed in that it aims at a relatively firm statement about the effects of a specified change: for example, the influence of a major industrial development on employment, transportation, demand for consumer goods, emission of pollutants and so on, all of which *might* be predictable from theory and previous empirical research.

Normative and extrapolative methods can be illus-

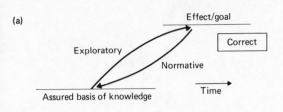

(a)

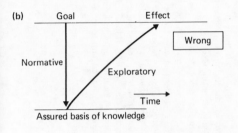

(b)

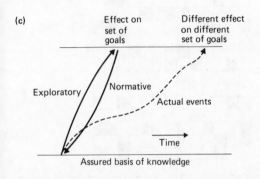

(c)

Fig. 7.4 Correct and incorrect feedback relationships between exploratory (extrapolative) and normative forecasting (Chadwick 1971)

trated (Fig. 7.4) with reference to Jantsch's (1967) statement (related to technological forecasting) which also forms the basis of Chadwick's discussion (1971). (Note that in their work they use the term 'exploratory' where we have preferred 'extrapolative'.) Normative methods begin with a statement of need (or a goal) and then seek to identify how it might be achieved. According to Jantsch and Chadwick, in the 'correct' procedure this need or goal is projected to some future time. Herein lies a fundamental difference between technological and social forecasting. In the former, we can reasonably assert, the identification of future needs is relatively simple: in the latter it may not be impossible but it is likely to be far more difficult to define needs with the same degree of detail and precision. Having established a set of

needs, the aim is then to reconcile them against present conditions and anticipated trends (i.e. involving extrapolative forecasting) and *also* to identify and solve the bottlenecks, technical deficiencies and other constraints that might be foreseen. Again one can discern a difference between technological and social forecasting. In the former, the reconciliation process is one that raises technological problems that may or may not be solvable within the time available. In the latter, reconciliation is a much more open, political process. However, in both cases it is possible to make productive use of a goal-compatibility matrix (Fig. 7.5) in which relevant goals and constraints are identified, together with the degree to which the pursuit of one is helped or hindered by the pursuit of others. In this example, the overall goal concerns the redevelopment of a hypothetical urban centre. A number of contingent goals are identified, and it can be seen that many of them are incompatible as things stand (Roberts 1974).

Extrapolative forecasting is now a familiar procedure, and is one where a large variety of techniques exists. At its most simple (though least acceptable) it is merely a case of extending a trend to some future date. The dangers are self-evident, since the shape of a trend depends on its determinants and these may not be stable. Hence extrapolation should proceed only on the basis of expectations founded on theory. This has long been accepted in population forecasting, in which fertility and survival rates play a crucial role but are insufficient without some knowledge of the mechanisms and determinants of migration (Willis 1974). Other forms of extrapolation include the use of 'envelope curves' and scenario writing. The first, which is really relevant only to technological forecasting, works on the principle that a sequence of innovations can be contained within an enveloping (or composite) trend (see Fig. 7.6). Each individual component exhibits exponential growth followed by a flattening of its curve, at which point it is superseded by another innovation. There are, of course, traps in using envelope curves. Firstly, it is easy to make a spurious extrapolation on a false expectation of the inevitability of progress. Secondly, these curves say nothing about *when* one should start to think about alternative technologies. Scenario writing has earned some disrepute by being used in an excessively grandiose manner (Kahn and Wiener 1967), but it is a technique that has considerable value in exploring the implications of alternative events and trends. It proceeds by identifying the 'switching points' that might lead to different trends

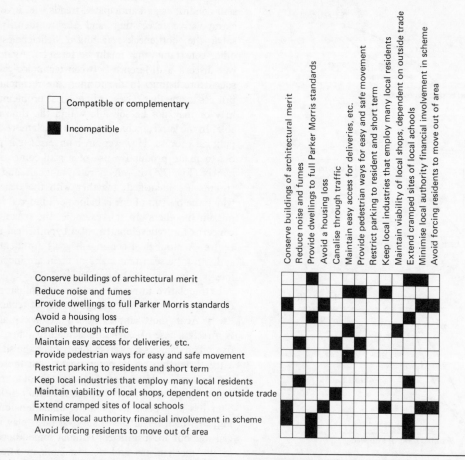

Fig. 7.5 A goal compatibility/conflict matrix for a hypothetical urban centre redevelopment scheme (Roberts 1974)

being established. Thus in several planning exercises the approach has been to examine what might happen if different sets of constraints were established (e.g. strong versus weak environmental protection) or if different levels of investment were carried out (e.g. a major road programme versus no additional routes or increase in capacity).

The other categories of forecasting methods mentioned above are systematic and intuitive. The first of these has already received attention indirectly, by noting the requirements for theory-based forecasting in which projections are determined from the interrelationships of all relevant factors. Most of the important work in forecasting follows this line and examples are found in research towards the refinement and operationalisation of (for example) input–output techniques in regional economies (Morrison 1974) or models in land use and transportation (Wilson 1974; Batty 1976).

Intuitive forecasting techniques deserve a little attention because there is a danger of their returning to fashion in a guise that pretends to be rigorous. The most common, known as the Delphi technique, was developed at the Rand Institute as a method of tapping 'expert' opinion through questionnaires, identifying areas of agreement and disagreement, and then asking for 'second-round' opinions to see whether greater consensus could be achieved (Enzer 1970). Superficially there is little to take exception to in this: after all, experts should be able to make reasonable judgements in the areas of their expertise. However, Sackman (1975) presents a powerful critique of the method as it has been applied, particularly in the United States. Most importantly he sees it falling far below the acceptable standards of social science research: it overlooks the possibility that consensus may in fact reflect subconscious peer-group pressure; experts tend not to be randomly drawn,

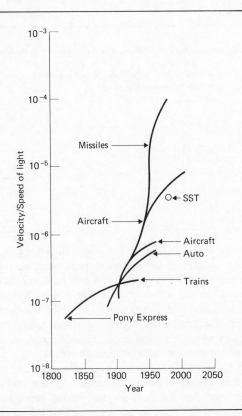

Fig. 7.6 Envelope curves in technological forecasting: the speed of travel

and so on. Sackman also cites a study which raises a question about the nature of expertise. Two groups of experts provided independent forecasts for 55 items on their expected timing of events. The result was an average variation between the two groups of 3.5 years for items that were expected to occur within the next 20 years. Thus, while there may have been consensus for some events, there must also have been almost total disagreement about the likely timing of others. With problems such as these it is therefore essential that claims of bogus authenticity are avoided and, where a Delphi technique has been used (East Anglia Regional Strategy Team 1974; Ley and Anderson 1975), it is crucial to examine the basis and method of its use.

The questions of the timing of events and the pace and magnitude of change are crucial if forecasting is to have any practical value. In a normative situation, where the forecaster may also exercise control over events, some of these problems may not occur. Having established an objective, the procedure is one of arranging events in such a way that the objective is

achieved by a due date. This is a task for operations research, and while it may involve a complex system, it is one from which much indeterminacy is excluded. But in more open systems, where the aim is to extrapolate trends or the impact of particular events it is difficult, if not impossible, to impose a deterministic framework. Moreover, while it may be possible to estimate a range of indeterminacy it is usually not possible to calculate a set of objective probabilities (see p. 76 above). The outcome is that in extrapolative forecasting the best that one can do is to identify a range within which likely conditions will be contained. This range will be defined in terms of both the magnitude of change and its possible timing. For most purposes the range will be bigger (and the indeterminacy greater) the further one looks into the future.

The first implication that might be drawn from this is that one should avoid making forecasts of the distant future – say, more than a couple of decades. However, two points can be raised here. One is that different phenomena have different forecasting horizons. Thus, for example, we can probably have greater confidence in a ten-year forecast of population than a ten-year forecast of demand for ornamental plastic gnomes. More importantly, our ten-year forecast is likely to attract greater confidence for national population trends than for regional or local trends where a more 'open' system applies. The definition of appropriate forecasting horizons has received very little attention but is clearly a problem that deserves study. The other point is that many decisions in urban and regional development do have a considerable impact over a long period. Chadwick (1971) has illustrated some of these (Table 7.3). Thus, in the case of major public utilities, it may take a decade from decision to completion, and the potential useful life may be several more decades. So, whether we like it or not, a demand for long-term forecasting cannot be avoided.

Forecasting and geography

Few geographers have been deeply involved in forecasting. This is true even of those who have been heavily involved in quantitative analysis and mathematical modelling. Most of their effort has been concerned with static patterns or comparative statics. Few studies have incorporated *time* as an explanatory

Table 7.3 Broad time scales in planning and development (Chadwick 1971)

Times in years, based on past experience:
Column 1 = Time required for planning, land acquisition and administrative procedures *before* site preparation and construction can start.
Column 2 = General construction time. This may be a function size and complexity of development to some extent.
Column 3 = Potential useful life of development, in terms of activities for which space initially adapted.

	1	2	3
Residential and general development	2	2	50–100
Town centre redevelopment	4	3–5	20+
Large-scale private development of new towns	5	10+	50–100
Motorways, major public utilities	8	2–4	20+

Suggested planning time scales	
'Intermediate'	0– 5 years
Middle range	5–15 years
Long range	15–30 years

variable in the analysis of change, and without this it is difficult to develop a proper framework for any form of forecasting. However, there have been several significant attempts to apply developments in theory and technique in this direction (Bourne 1974) and there will doubtless be more, particularly in the field of impact forecasting which is a necessary input for economic evaluation and environmental impact assessment.

Extrapolation has a long history in social and economic analysis and is well established in the form of population forecasting. Demographic change is important in its own right but the role of population as a determinant of other conditions gives it added significance. Housing, social services, and consumer demand are directly affected and some types of travel, employment and pollution are indirectly influenced. Population forecasting also illustrates the important distinction between endogenous and exogenous factors. A dynamic element exists in all population structures which have the ability to reproduce themselves. If there were no external forces this would allow a population projection to be made solely on the basis of the existing age distribution together with cohort survival rates and fertility rates. However, exogenous factors have two influences. Firstly, they can affect the internal dynamics: survival rates and fertility can be influenced by medical advance and changing lifestyles, respectively. Secondly, external forces operate through the process of migration which, in turn, is influenced by a wide range of economic and social factors (Willis 1974; Godlund 1971; Thompson 1971).

At this point what may start as a problem in extrapolation begins to take on some of the features of impact forecasting. That is, in addition to knowing

the internal dynamic of a demographic system one needs to predict the impact upon it of changes in exogenous variables – which also may need to be predicted. At the urban and regional scales these include factors which are common to a wider environment (e.g. medical and technological innovation) and factors which are localised (e.g. access to economic opportunities and local planning objectives). In practice these are often assumed to be constant, or to fall within a range of possible conditions, and a projection made on the basis of the internal population dynamic. However, as one moves down the spatial scale this approach is increasingly put under pressure by the variability of these factors.

Geographically based, socially disaggregated population forecasts have a wide range of uses and are a major input for most types of planning. For example, demand for accommodation in an area depends largely on the size and number of households and rate of household formation. Meeting that demand requires land and other building resources, or a replacement of existing residences by new ones at a higher density or, a third possibility, diverting the population growth to new centres of development (Roberts 1974). At the same time it is possible to extrapolate the trends and (given other exogenous information) the geographic distributions of certain social or cultural groups, with the aim of developing appropriate policies to meet their specific needs (e.g. children and schools; the elderly and social services).

The probabilistic nature of extrapolative forecasting is shown by a technique that has attracted considerable attention in recent years – namely, *Markov-chain* analysis. Here the aim is to predict the likelihood of an object being in a certain state at a certain time, given these characteristics at a previous time. An initial assumption needs to be made that the patterns observed in the immediate past reflect probabilities of transition from one state to another which will remain the same in the future. (This assumption may later be relaxed for experimental purposes.) The technique is useful in exploring possible trends in the distribution of change and has been used in a variety of studies dealing, for example, with land-use conversion (Drewett 1971), industrial movement (Lever 1972), household migration (Collins 1975) and ghetto formation (Berry 1971). An example, relating to the relocation of factories among zones in the Glasgow region, is shown in Table 7.4 and Fig. 7.7.

Table 7.4 Markov-chain analysis of industrial movement in Glasgow (Lever, 1972)

To		Zone 1	Zone 2	Zone 3	Zone 4
	Zone 1	118	13	4	14
From	Zone 2	6	33	8	6
	Zone 3	1	1	68	5
	Zone 4	2	0	3	43
		Zone 1	Zone 2	Zone 3	Zone 4
	Zone 1	0.79	0.09	0.03	0.09
	Zone 2	0.11	0.63	0.15	0.11
$P =$	Zone 3	0.01	0.01	0.91	0.07
	Zone 4	0.04	0.00	0.06	0.09
		Zone 1	Zone 2	Zone 3	Zone 4
t_0 (1959)		46	16	23	15
t_1 (1969)		39	14	26	21
t_2		33	13	28	26
t_3		29	11	30	30
t_4		25	10	32	33
t_5		23	9	33	35
t_6		21	8	34	37

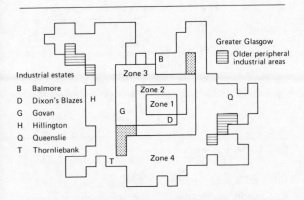

Fig. 7.7 Zones used in a Markov-chain analysis of industrial movement in Glasgow (Lever 1972)

The last aspect of extrapolative forecasting to be dealt with here is one that is relatively new in its methodological treatment, but which has always been a focus of interest. This is the prediction of the point at which a trend will peter out, or even collapse, and of the conditions which will succeed it. Formally, this has been approached through the mathematics of *catastrophe theory* (although the mathematical content is less important than the conceptual framework) which deals with discontinuities in otherwise continuous systems (Thom 1975). It is quite easy to think of physical examples: the steady build-up of destabilising conditions leading to landslips (Brunsden and Thornes 1979) or the accretion of combustible vegetation prior to major bush-fires (Luke and McArthur 1978). The human realm is far less easy to predict in this way. Even the interpretation of historical events in terms of catastrophe theory has been treated sceptically (Wagstaffe 1978; Baker 1979), so it may be over-optimistic to expect much from its formal use in prediction. Yet we continuously allude to it in everyday life whenever trends appear destined to end in conflicts either among groups of people or between people and their environments. If catastrophe theory only formalised what is already being done informally it would be of little value. However, at the very least, it appears to offer a useful framework examining the nature and possible outcome of trends.

In comparison with extrapolation, impact forecasting adds another dimension in that it is concerned not only with the internal dynamic of a system but also with the relationship between one system and another. That is, a change in one causes an impact on the second. They may both be human systems (e.g. the effect of new population on land use and transportation in a region) or one may be a human system and the other an environmental one (e.g. industrial growth and effluent disposal on a river system). Generally it seems true to say that in both cases emphasis has tended to lie on the system being affected. This is reasonable, as long as the system and its external relationships are treated fully. However, it also appears to be true that most impact forecasting treats the impact as a once-and-for-all event which initiates a chain reaction in the affected system, rather than a more complex interaction between systems.

Impact forecasting in urban and regional development has concentrated on two sets of characteristics: accessibility and activities. Each can be treated independently, or partially, but they have also been combined in general forecasting models of land use and transportation (Wilson 1974; Lee, C. 1973; Batty 1976).

The modern origins of these large-scale models are found in practical planning exercises in the USA during the 1960s, at a time when the growing urban crisis was forcing policy makers to review the use and distribution of resources. At this time a wide range of models was devised whose aim was to predict the impact of alternative actions and, thence, to find optimal plans. Many of them had weak theoretical underpinnings and many also proved to be over-ambitious in terms of data and computer requirements. More recent experiments have partly rectified the theoretical deficiencies and, at the same time, have benefited from developments in computer systems.

One model in particular has a long-lasting influence. This is the Lowry model, which was developed in 1963 in conjunction with the Pittsburgh Community renewal project (Lowry 1964). It has subsequently been refined and extended in a number of ways but has retained its basic structure and problems (Bernstein and Mellon 1978). Part of its attraction has been its foundation upon widely known concepts of spatial structure and economic activity: the gravity model and economic base theory (see Chapter 5 above). The model uses three sub-systems, employment, population and transportation, and aims to allocate to geographic zones the population and non-basic employment and, by extension, the pattern of journeys of various kinds, consequent upon a change in basic employment.

Having identified and collected its relevant impacts, the model works through an iterative sequence of steps as described in Fig. 7.8. Here its starting point is the creation of new basic employment, located in a particular zone: this might be a new industrial complex. Each job is assumed to be filled by one member of a household, and the average size of households can be used to estimate the total population supported by the new basic employment. Subsequently, the new population creates a demand for service (non-basic) activities, the size of which is estimated from previously established population/service ratios. Meanwhile, population and activities are allocated deterministically among zones using the gravity model, together with such constraints as the amount of usable land available in each zone and the market thresholds necessary for an activity to locate in a zone.

A number of points have been made in criticism of

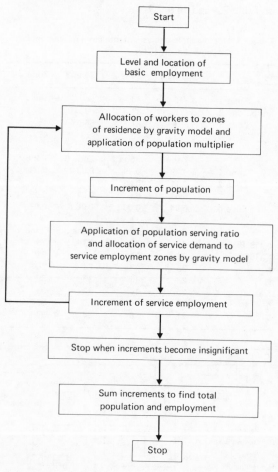

Listing of variables

Exogenous

1. Sub-area distribution of employment used by the export industry.
2. Amount of usable land in each sub-area.
3. Space occupied by export industry in each sub-area.
4. Retail production functions: fixed amounts of space and labour required per customer in each retail line.
5. Air distances between sub-areas.
6. Labour force participation rates.

Endogenous:

7. Population by sub-areas.
8. Sub-area total employment.

Fig. 7.8 The structure of the Lowry model

this and similar models. It ignores many variables that might be significant: unemployment, property values, government policies, social discrimination, and so on (Bernstein and Mellon 1978). It is also a model that relies heavily on consumer demands – expressed in demand for services and in the disutility of travel – as the means for allocating activities and movements. Also, it treats these demands deterministically, using the parameters of previously established gravity models. (The last criticism stimulated a methodological refinement in the introduction of entropy-maximising principles, whereby activities and movements are allocated stochastically.) Lastly, it uses economic-base theory in a form that, even before Lowry devised the model, had been shown to have serious shortcomings (Massey 1973). What these criticisms amount to is a statement that the greatest danger in forecasting is not over-elaboration as some critics have argued (Lee, D. 1973) but over-simplification.

8 Evaluation and assessment

Decision making involves a choice between different actions, or between action and inaction. In either case there is a need to evaluate the alternatives. Sometimes a choice is obvious and immediate; at other times it is necessary to balance a range of advantages and disadvantages. Ideally the use of an evaluation technique will point unambiguously towards one of the alternatives, to the exclusion of all others.

When an action or project is proposed it needs, eventually, to be subjected to two questions: is it feasible? and is it worth-while? Feasibility is largely a matter of technology, resources and organisation which together determine the ability of a community to implement its proposals. Whether or not something is worth-while depends on how much the community, or some member of it, will benefit. This may be a clearly identifiable economic benefit, but it need not be; it could justifiably be a cultural or environmental benefit for which there is no direct economic equivalent.

Whether or not something is worthwhile presents a more complex set of issues than the question of feasibility. While the latter may need considerable investigation and, even then, only be answered in conditional terms (i.e. 'project x would be feasible if . . .') the answer has a certain objectivity. The notion of 'worth-while', on the other hand, raises two crucial questions which, however hard we try, we often cannot answer totally objectively. These are: worth-while by what criteria? and worth-while to whom?

Thus, the achievement of soundly based evaluations is not always a simple task. Firstly, there is a basic question of ensuring that the concept and content of an evaluation technique meets the requirements of a plan or project that is to be examined. Secondly, it requires an effective information system which, as has already been discussed, is not always automatically forthcoming. That there are now many different evaluation techniques is testimony to the stringency of these requirements. Each differs from the others in the way in which it handles the various aspects of evaluation and in its comprehensiveness (Lichfield 1970). Each also differs in the degree to which it has been tried and tested, and it is safe to assert that this is an area which will continue to repay considerable attention for some time to come.

The focus of this chapter is less on techniques (though these are not overlooked) than upon the uses and potential uses of evaluation methods and the contribution that geographers can make in their application. Every decision about the use and distribution of resources is a choice among alternatives. Our inhabited landscapes – and many uninhabited ones, too – look and function the way they do because of choices made by individuals or communities.

But there is not, and never can be, one 'right' way to evaluate alternatives, and this is reflected in the present chapter. Firstly, we follow convention by discriminating between economic and environmental evaluation; secondly, we look at different techniques within these categories. This is not a wholly satisfactory approach, because economic and environmental evaluation methodologies overlap in several ways; but it is a convenient way of proceeding.

The starting point is cost–benefit analysis (CBA). More simple techniques could have been introduced first, but there is value in going directly to the technique which is most demanding in conceptual and practical terms and which, also, contains the logical bases for many others. By first examining CBA we have a foundation for assessing other techniques for both economic evaluation (such as cost-effectiveness and planning balance sheets) and environmental impact evaluation.

Lastly, throughout the following discussion it will appear as if evaluation and assessment techniques come into use *after* a project has been proposed or a set of alternatives has been defined. However, while this is where they have their greatest importance, all of the techniques described also play a part in the *design* of plans and projects.

Cost–benefit analysis

The net advantages of an innovation or investment can accrue to an individual as 'profits' or to a society as an increase in welfare. In CBA our interest lies with the latter. Profit is a concept implying competition and the appropriation of benefits by an individual for himself: it may also imply a reduction in profit, or a loss, for other individuals. It has a place in economic geography, in normative models of location, where it is an indicator of a firm's locational efficiency (Losch 1954), but not in the measurement of the general welfare of society. Here, net social benefit is derived by summing all benefits regardless of to whomever they accrue, and subtracting all costs. Also, it is unimportant whether these costs and benefits are the direct or indirect outcome of the innovation or investment. As long as they can be attributed to it they form part of its welfare effect.

This has proved easier to conceptualise than to implement. Costs and benefits are often obscure, and even when they are identified it may be difficult to assign values to them. Anything that is ignored is implicitly given a zero value. Then, since costs and benefits often occur as a stream continuing into the future there are the questions of forecasting their magnitudes and assessing their worth in *present* values. And, following the earlier discussion of uncertainty, these forecasts may only be best estimates of future conditions. Somehow, an evaluation needs to take account of risks attached to a project.

Many questions will always appear insuperable, and this will remain a source of discontent with CBA. The same can be said of other techniques. What CBA offers, however, is a logical framework for analysis (Mishan 1972; Dasgupta and Pearce 1972; Layard 1972; Newton 1972).

The primary aim of CBA is to estimate the *present net value* of a project. Where this is positive the project will contribute to aggregate welfare (i.e. the benefits outweigh the costs) and vice versa. However, it may not be enough to maximise net present value if, in so doing, some people bear most of the costs and others reap most of the benefits. Thus, some criterion of acceptability is necessary. One criterion used

in welfare economics is the concept of Pareto optimality, which states that a policy is acceptable if at least one person prefers the new situation and no one loses by it. If this were applied in the real world then very little would ever get done. Hence the general adoption in CBA of a more liberal formula, the Hicks–Kaldor criterion, which defines acceptability as a state in which the gainers could compensate the losers, even if they do not actually do so. Effectively this compensation principle is a restatement of the Pareto criterion, altered only by a recognition of the high probability of unequal gains and losses. Its significance has been widespread: it is found in much planning legislation, particularly in relation to compensation and betterment arrangements, and in the handling of external disbenefits arising from public projects (e.g. noise from airports).

In summary, the information required to conduct a CBA falls under the following:

valuation of each cost and benefit at the time when it occurs;
discount rates to be used in translating future values to present values;
estimates of risks and their effects; and
the distributional effects of the project.

Before discussing these in a little detail we can use a simple, hypothetical example (Fig. 8.1) to describe how they are used.

Imagine an agricultural district in which there is just sufficient precipitation to support marginal farming operations. Each farmer is able to increase output by boring wells and extracting groundwater for irrigation at an average cost of 5¢ per kilolitre. Demand is such that, at this price, 50,000 kilolitres are consumed each year and that if the price were reduced consumption would increase up to a maximum of 100,000 kilolitres, at which point extra water would be detrimental. Thus the farmers are paying $2,500 per year for borehole water.

Now assume that the construction of a dam and irrigation system is proposed which will have a once-and-for-all cost of $40,000. Miraculously it has no upkeep costs. In addition, the proposal is to supply to the farmers, free of charge, 100,000 kilolitres per year, and there are no risks of this figure not being reached, nor of any problems occuring in the system. Nor are there any side-effects on non-farmers or on residents of other regions. The scheme is intended to last for ever.

Should the project be carried out? To answer this

(a)

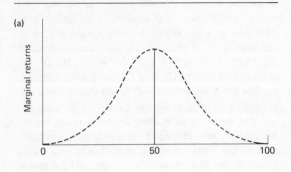

(b)

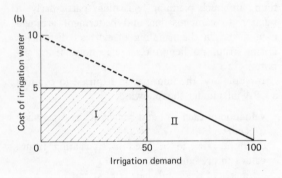

Fig. 8.1 Measuring benefits in cost–benefit analysis: a hypothetical example

we need some additional information. First, we need to know the full value of the free irrigation water to the farmer. We know that the 50,000 kilolitres of borehole water which cost $2,500 is now supplied free, so that is a benefit from the scheme. But what about the extra 50,000 kilolitres that is now consumed? At this point we need to know the shape of the demand curve for water; that is, the amount that farmers would have been willing to pay for it. We might be able to estimate this from surveys, but here we shall assume that kilolitre number 50,001 would have been consumed at a price just under 5 cents and that kilolitre number 100,000 will only be consumed if it is free, and that the demand curve between these two points is a straight line (Fig. 8.1(b)). Because this is what farmers would have been willing to pay we can infer that the benefit of having the extra 50,000 kilolitres amounts to $1,250. So the total monetary value comes to $3,750 per annum.

If the value of a fixed sum of money is constant regardless of when it is received we could say that by the end of the eleventh year of operation the benefits of the project would have exceeded the costs and the project should go ahead. However, common sense

(and observation) tells us that this is not so. Given the choice between $1 now and $1 in one year's time we should prefer the former. Just how much we prefer it is expressed in a discount rate. A rate of 10 per cent means that we would be indifferent between $1 now and $1.10 in a year's time. Or, to put it another way, that $1 next year is worth only 91 cents to us seen from today's viewpoint; the following year's dollar is worth only 82 cents, and so on. Clearly the choice of discount rate can be crucial. Too high a rate may mean that otherwise worthwhile projects will be neglected; too low a rate may mean over-investment.

To illustrate this, Table 8.1 sets out the costs and benefits using three different rates. At 1 per cent the project is clearly worth-while, yielding a net present value of +$335,000, while at 9 per cent it is just worth while and at 10 per cent the costs exceed the discounted benefits.

Table 8.1 Net present value of an irrigation scheme.

	Future net benefits per year forever	Preset value at discount rate of		
		1%	9%	10%
Existing users	+$2,500	+$250,000	+$27,778	+$25,000
New users	+$1,250	+$125,000	+$13,889	+$12,500
Taxpayers	–	–$ 40,000	–$40,000	–$40,000
Net present value		+$335,000	+$ 1,667	–$ 2,500

Note to calculate n.p.v. at a given d.r. forever, the formula is P/r, where P is annual benefit and r is the discount rate expressed as a fraction (ie. 10% = 0.10).

There remains the question of distributional effects. If the project is paid for by taxpayers then the farmers will, of course, always promote it, whatever the net present value to society (getting something for nothing is the essence of politics), but it will certainly not be in society's interest to support it unless there is a positive net return. However, this is not the whole story. If the farmers happen to be much richer than the rest of society, it is likely that their utility of each dollar of extra benefit is less than the disutility of each equivalent dollar of cost paid by other taxpayers. If this is so, then the net present value needed to justify the project will have to be adjusted upwards.

It would be rare to find anything as simple as this hypothetical case. Most public projects have many more direct and indirect costs and benefits to deal with and also have to come to terms with different ways of evaluating them. The following paragraphs discuss some of the issues involved.

The first requirement listed above is a definition of what costs and benefits to include which, in turn, requires an accurate definition of the system affected by the project. Direct costs involved in implementing the project ought to pose few problems, but indirect ones, as well as benefits of both kinds, may be more difficult. For example, an urban redevelopment scheme might influence custom in another city; a new airport might attract new travellers; or an oil terminal might adversely affect tourism and fishing. Defining the system therefore requires, first, identifying the individual items affected and, second, making a cut-off between what is significantly affected and what is not. The definition should be both sectoral and geographical. It should also be temporal, so that values can be imputed over an expected time span defined by the life of the project.

Despite the requirements that *all* costs and benefits should be included there is, as Mishan (1972) has argued, a likelihood that a degree of asymmetry will occur. The act of drawing lines between what is, and what is not, to be included means that some items, however insignificant, will be given a zero value. This could be serious if the limits are too narrowly defined. For example, Mishan quotes the use of the Noise and Number Index (NNI) by the Third London Airport inquiry, arguing that the line used as a cut-off (NNI = 35) excluded a large number of people who incurred a disamenity from noise.

Asymmetry may also occur because individual costs and benefits are unequally easy to measure. Some items can be estimated from market values; some are not marketed but do appear in the market value of others; some have no market values at all. The first group is likely to include the capital costs of a project and many of the operating costs, although all of these can be distorted by various conditions, such as underemployment, tariffs or subsidies. The second includes items such as the disamenity of noise or pollution, or the value of accessibility, which may appear in property values. The third contains various 'intangible' qualities of objects and lifestyles. The example of historic buildings valued at their insured value inspired Adams (1970) to suggest that London's Third Airport should be built in Hyde Park where the benefits of proximity to the city would far outweigh the cost of a few old buildings in Westminister and Kensington.

Where markets are inadequate or inappropriate, it is necessary to estimate what people would be willing to pay for a good, or would accept as compensation for a disamenity. Immediately this raises problems, as discussed earlier in Chapter 3. Estimates based on consumer behaviour are likely to contain biases: for example, travel costs may ignore the costs of pre-trip planning. The use of CBA in environmental evaluation has uncovered a number of other issues. Krutilla and Fisher (1975) have identified 'option' and 'existence' values in addition to *exhibited* willingness to pay. Option value occurs among people who are not currently using a resource (e.g. a national park) but who might do so at some future time and would therefore be prepared to keep their option open by paying towards its preservation but may be denied the opportunity to do so. Existence value occurs among those who do not intend to use the resource but who derive satisfaction simply from the knowledge that it exists. The reasons of both groups may be rational (i.e. the resource has scientific or educational value) or eccentric, but that is not important as long as they contribute in terms of willingness to pay. To ignore them is to bias the evaluation.

The second information requirement – determining the proper rate of discount – is one where geographers have little to contribute. Even economists appear to be uncertain about the proper method of calculating rates and much discussion has been conducted on the relative virtues of time preference rates (i.e. the degree to which present consumption is preferred over future consumption) and the social-opportunity cost of capital (i.e. the rate of return that could be earned in a different use, usually assumed as being in the private sector). The arguments for and against these measures are summarised in Dasgupta and Pearce (1972) and Layard (1972). Whichever method is used, the crucial point is the identification of the right rate, since even small errors can have a major effect on the outcome, especially where the project is intended to have a long life (O'Riordan 1976). If benefits are discounted at too high a rate this will tend to favour projects which yield a more rapid rate of return and whose significance may be felt in both urban and environmental planning. Compare the hypothetical figures in Table 8.2. Option 2 has a much greater advantage over Option 1 when discount rates are high than when they are low.

Uncertainty and risk, which were discussed in Chapter 6, play an important role in CBA. Projection of costs and benefits is, at best, only feasible in

Table 8.2 Discount rates, project life and the evaluation of costs and benefits

Year	1	2	3	4	5	Total
Option 1 flow of benefit	0	100	100	100	100	400
discounted at 10%	0	83	75	68	62	288
at 2%	0	96	94	92	91	373
Option 2 flow of benefit	400	0	0	0	0	400
discounted at 10%	360	0	0	0	0	360
at 2%	392	0	0	0	0	392

probabilistic terms with upper and lower limits of confidence surrounding a best estimate. Furthermore, each cost and benefit has its own set of probabilities. Thus an early task in CBA is to identify and estimate the various risks attendant upon a project. For example, in our hypothetical case in Figure 8.1, the cost of the irrigation scheme would be likely to range between just above and just below $40 m. The stream of benefits, which are projected much further into the future, are likely to show a wider range of probabilities. We could also have complicated matters by allowing for the effects of environmental variability and fluctuations in the price of agricultural products.

Once risks have been identified, they can be incorporated into the analysis. Broadly there are two main ways in which this can be done (Arrow and Lind 1970; Layard 1972). One is to experiment with the various estimates of every cost and benefit, together with their probabilities of occurrence. The other is to work with the best estimate and to vary discount rates by adding a premium to the rate for more risky items. In either case the outcome is a review of the sensitivity of the project's value under different sets of conditions.

The application of these measures, and the response to their outcomes, is also related to the character of a decision maker. The more risk-averse he is, the more he is likely to devalue benefits and inflate costs. However, the same logic has led some economists to argue that in some public investments, where costs and benefits are defrayed among millions of taxpayers, the effects of risk tend towards zero and that, therefore, there is no need to make adjustments (Arrow and Lind 1970). Exceptions to this might occur when costs can be defrayed widely, but benefits accrue to a small section of the public or vice versa. In these cases it might be appropriate to adjust for risk.

Lastly, we come to the thorniest problem of CBA – the assessment of distributional effects. It is here that the most damaging criticisms have been made (Broadbent 1977; Smith 1977). The major issue concerns how the utility of each dollar of costs or benefits accruing to people with different incomes is assessed. The easy way out is to assume that everyone – rich and poor alike – has the same set of utilities. However, to do this may be to bias severely the outcome, especially if costs and benefits fall unequally on different groups, and, thereby, to redistribute real income from one group to the other. Layard (1972) has insisted that equal utilities will only occur by chance and that 'for any reasonable welfare function they will be higher for the poor than the rich'.

So the question of distribution effects has two aspects: how to define the utilities of different groups and how to act upon the information once it has been obtained. Neither can be answered clearly. For the first, we might rely on the assumption that the marginal utility of income is inversely proportional to an individual's post-tax income; or we could try to refine the answer by surveys. But whatever method is used there is, we can speculate, a role which can be fulfilled by geographers through the extension of their work on regional disparities. This has tended to be limited to the description and explanation of inequality (e.g. Sant 1974; Coates, Johnston and Knox 1977). By associating it more closely with the concepts of welfare economics in general and CBA in particular, it might be possible to carry out more discriminating analyses of: (a) the impact of projects on income distribution; and (b) the effects of income distribution on the evaluation of projects.

Assuming that utilities can be defined, the question then is to define what is an acceptable change in welfare arising out of a project. Smith (1977) has approached this by presenting graphically (Fig. 8.2) a set of ideological views of distributional acceptability. These range from a Benthamite attitude that any net increase is preferable regardless of effect on distribution, through a series of increasingly stringent constraints on the meaning of acceptability. Although these embrace the range of attitudes towards equality found in a society, in reality it is more realistic to expect attitudes to be associated with activities. For example, we would probably be concerned that social-service provision conforms more closely to principles of equity than, say, the regional development impact of a new industry.

We have tried, as far as possible, to give a balanced account of CBA. At the same time we have to

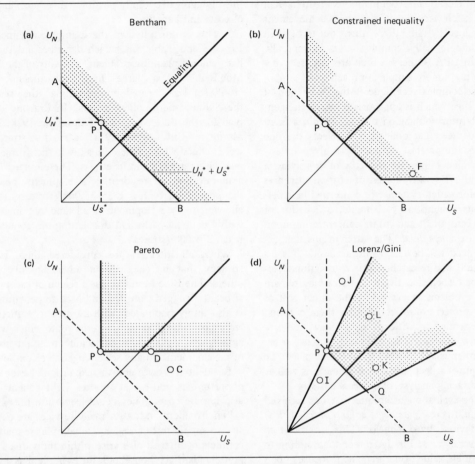

Fig. 8.2 Alternative criteria for judging improvements in distributions among regions (Shaded areas show welfare improvements) (from Smith 1977)

admit that, on a practical level, it is fraught with difficulties. Information is often not available, questionable assumptions are used, it is impossible to be all-inclusive in the identification of costs and benefits, and we have to make judgements about an uncertain future. But, then, what technique is free from these problems? The more serious issues are conceptual and concern the question raised by Broadbent (1977) of whether the micro-economic welfare theory underlying CBA is a valid description of what happens in urban and regional systems. That is, while the logic of CBA might be internally consistent, the dynamic life of cities often is not. Social groups change and, over time, so also do preferences and utilities; consequently it is often difficult to say who is better off.

Yet despite these criticisms CBA represents a set of principles which underlie not only economic eva-

luation but many other forms of assessment as well. It may not be possible to put monetary values on everything, and CBA is not readily applicable to major, highly complex schemes with many external effects. And it may be that the results contain a bias due to the asymmetry of costs and benefits. But, for all that, it is the most valuable starting point that we have in analysing the desirability of actions.

Alternative economic evaluation techniques

It is not obligatory to use CBA, and other techniques are available. Some have been designed deliberately to escape from problems which arise in CBA although, in fact, this freedom is more apparent than real. One such technique, the goals-achievement matrix designed by Hill (1973), turns out to be even more methodologically complex and conceptually suspect than CBA. Others, which are discussed below, try to use a more simple approach.

The least complex technique is the estimation of *cost-effectiveness* which is appropriate where an agency has a statutory obligation to meet within a fixed budget. To meet that obligation it selects the one which gives the greatest output for a given investment, or which minimises the cost of supplying a given output. The advantages of cost-effectiveness as far as information systems are concerned lie chiefly in: (a) its assumption of benefits, which do not have to be computed; and (b) its concentration upon costs that are *internalised*. The latter means that externalities need not be taken into account, thus reducing greatly the research required to estimate social costs or disbenefits. Even so, there may remain significant problems in evaluating the direct cost of alternative projects or of forecasting their outputs. However, while cost-effectiveness may indicate which of two alternatives is preferable, it does *not* say whether either is worth-while in the first place. To do this requires a deep analysis of benefits as well as a more embracing analysis of external costs.

The other evaluation technique that has received wide attention is *planning balance-sheet* (PBS) analysis, devised by Lichfield. This differs from CBA in several ways. Firstly, rather than aiming to provide a single value of net benefit, it focuses upon the distribution of costs and benefits among the various sectors affected by a plan. Secondly, it allows non-monetary measures where these are the only feasible means of comparing alternatives. In some instances where the technique has been applied these have outnumbered the monetary measures (Lichfield 1969; Lichfield and Chapman 1968). Thirdly, PBS is claimed to be suitable for complex projects involving many different sectors and agencies. This is made possible by the measurement of costs and benefits in relation to the objectives of each of the groups affected.

Although it has been described as an alternative method, Lichfield has emphasised that PBS is essentially derived from CBA and, in many respects, uses the same procedures and concepts. Where it differs is in its aim for maximum pragmatism by adopting a 'second-best' approach over issues that would otherwise be too difficult to resolve, and by emphasising the disaggregated (and, possibly, divergent) nature of costs and benefits.

This is illustrated in the example contained in Fig. 8.3 and Table 8.3, which describes an Australian regional planning problem. The City of the Blue Mountains is a large local government area (1,400 sq. km) located immediately to the west of the Sydney metropolitan region. In drawing up a plan for the Blue Mountains (Alexander 1978; Blue Mountains City Council 1974), alternative strategies were defined which differed mainly in their projected population growth and its effects. These ranged from a moratorium on residential development; holding the population to less than 55,000 (strategy 1), to allowing full development of all land not included within national parks and thus letting the population reach 275,000 (strategy 5).

When all interests are considered there is no strategy that has a clear-cut advantage over the others. The Blue Mountains is a region of exceptional beauty: rugged, forested and hard to penetrate. It is also thinly populated, with most of its settlement strung out along the main arterial route which winds along ridge tops. In this environment the possibility of development, on any scale, engenders conflicts.

Residential growth and the expansion of economic opportunities and social services yield benefits but also impose costs. They could jeopardise the physical environment and, also, many services are expensive to supply in this region. To complicate matters, conflicts of interest also arise within individuals and households. Each household, for example, has a variety of wants: jobs, schools, shops, attractive residential areas, low rates and unblemished natural environments.

The balance sheet (Table 8.3B) attempts to account for all of these wants. Firstly, it identifies each group involved in or affected by the strategies and classifies them as producers and consumers. (The former are those who implement a plan.) Many of these groups overlap and many people are found in more than one group (e.g. ratepayers and residents). However, each group is distinguished by its objectives. Thus, in his capacity of land owner a person would (presumably) want to maximise his property value; the same person as shopper would want to maximise accessibility and choice. Secondly, it defines an appropriate measure to express the effect of each strategy on each objective. Herein lies the greatest contrast with CBA, which concentrates on a single objective and measure, namely money.

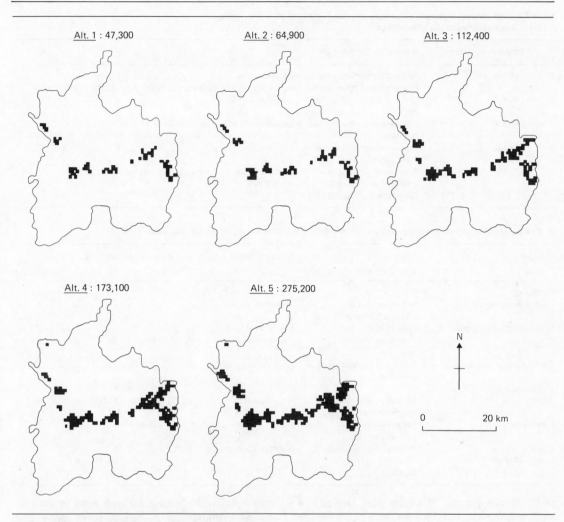

Fig. 8.3 Alternative strategies for the Blue Mountains, NSW

The next step is to estimate the 'costs' and 'benefits' of the alternative strategies. In this example only 2 out of 50 items have a monetary measure and 22 use some other form of quantitative estimates. The rest use qualitative assessments of the magnitude of costs and benefits.

Finally, these evaluations are subjected to a simple, unweighted ranking and addition of ranks (rank 1 = 1 point) to identify which is the preferred strategy. Without doubt this uncovers the Achilles' heel of PBS which no amount of manipulation (see, for example, Lichfield 1969) has been able to cure. As long as the great majority of items show a close concordance there is not much problem, but as soon as there is conflict the ordinal scale imposed by the

lack of a common unit of measurement creates the risk that the results will be meaningless as a guide to action. While some people find it difficult to face a column of figures without adding them up, this is an occasion when it is better not to do so. Instead, the ranked orders of preference on individual items could be used as the basis of discussion on the relative merits of the different strategies.

Another point of departure from conventional CBA is that the former, in theory at least, aims to say whether or not any action is worth-while. Here, as in cost-effectiveness analysis, the aim is just to find the most preferred strategy.

The analysis of the Blue Mountains, while not as thorough as it might have been, illustrates these

Table 8.3 Evaluation of alternative strategies for the City of the Blue Mountains
A. Main characteristics of the alternatives

	Target population	Allowable development
1.	47,300	Redevelopment of existing structures only. Undeveloped subdivisions may be required to conserve or restore to natural environment
2.	64,900	Development within existing serviced subdivisions. No flats
3.	112,400	Full development of existing subdivisions at present densities. No flats
4.	173,100	Full development at higher densities. Flats permitted
5.	275,200	Maximum development, including additional developable areas with slopes less than 1 in 5 adjacent or connected to serviced areas. Densities as in alternative 4

B. Extract of planning balance sheet analysis for the City of the Blue Mountains (after Alexander 1978)

Groups	Objectives	Measures used	Cost/benefits of strategies				
			1	2	3	4	5
Producers							
Dept of Main Roads	Minimise cost of efficient transport	New road works	0	1	1	2	2
Fire Protection Authority	Minimum free risk	Fire control cost	C_o	$C+$	$C+$	$C++$	$C+++$
Council	Provision of social facilities	Rateable income	B_o	$B+$	$B++$	$B+++$	$B+++$
Developers	Maximise development potential	New building lots	Nil	11,272	15,752	20,261	57,000
Consumers							
Youth	Improved employment opportunities	New local jobs	1,000	3,200	11,000	24,000	44,000
Landowners	Enhancement of property values		$B+++$	$B++$	$B++$	B_o	B_o
Shoppers	Improved local shopping	Number of centres	1	2	3	6	9
Holidaymakers	Maintain/improve facilities		$B+$	$B+$	$B++$	$B++$	B_o

points. Developmental strategies were favoured by local government and business interests while, in general, consumer interests were best served by conservationist strategies (except where these interests concerned improvement to services or job opportunities). Not surprisingly when divergence occurs, summation of ranked preferences leads to a compromise strategy being the one that scores most highly. The recommendation that was finally made was that the area should have a moderate level of growth to reach about 100,000 by the end of the century.

Another early example of multiple-objective evaluation, proposed by Hill (1968, 1973), is the *goals-achievement matrix*. As with PBS, a number of alternative strategies, and groups potentially affected by them, are identified. The technique differs, however, in that it attempts to make objectives as explicit as possible, and tests each group's costs and benefits against every objective. (In contrast PBS used a different objective for each party in the evaluation.) Although it is realistic to recognise that complex plans do have multiple objectives, Hill's technique has proved difficult to implement. His own experiments in Cambridge, where he re-evaluated the transportation proposals made in previous plans, contains much indeterminacy. Its use in regional planning, described in Chapter 9, has tended to be much looser than was originally conceived.

Environmental impact assessment

In turning to the effects of human activities on the physical environment we open up conceptual and practical problems which are at least as challenging as those posed by economic evaluation. The size of the challenge can be expressed in this way. To decide whether it is worth while changing a physical environment (which, in effect, usually means degrading it) we need to know three things: what the impact of the innovation will be on the environment, what the cost will be of protecting it, and what the economic advantage will be of using it. Collectively, these require a form of analysis which integrates both physical and social sciences and which allows each to gain from the interest of the other. Potential disturbance to an environmental system is an insufficient reason, by itself, to reject a proposed innovation, just as economic gain is not a sufficient reason to justify it (O'Riordan 1976; Greenberg, Anderson and Page 1978).

Left to themselves, individuals pursuing their own private economic interests have little incentive to protect environments except in cases where their own property values or conditions of life are concerned (Hardin 1968). This does not prevent some people from believing that their living conditions are affected by distant environments or others from using ethical arguments in support of conservation. But if economic values dominate then other values must inevitably be demoted. Combining the former with modern technology and the pursuit of scale economies, the potential for environmental damage – and for inflicting lower environmental quality on others – is greater than ever before.

These considerations, together with uncertainty about how environments respond to major developments, has created a demand for thorough appraisal of such schemes prior to their implementation. In the United States this led, in 1969, to the passage of the Environmental Protection Act which required an environmental impact statement (EIS) before any major federally supported development was carried out (Canter 1977). Some states followed the federal Act with legislation of their own, and a number of other countries have followed suit, and there have also been proposals to introduce international legislation (Holdgate and White 1977). In addition, the law has been extended to cover private as well as public development. However, it should be stated that there has also been reasoned opposition against the procedure in countries (e.g. the UK) where other planning legislation is already strongly established (O'Riordan and Hey 1976).

The circumstances in which an EIS is required are obviously of great importance. A definition which is too liberal will allow many actions which may have damaging effects to go unchallenged, while one that is too strict will be costly to implement and may have detrimental effects on economic welfare, leading to general opposition to environmental legislation. But finding the 'right' definition is not always straightforward. The American one, which has been adopted elsewhere, requires an EIS where a development is likely to 'significantly affect the quality of the human environment'. Without trying to split hairs, 'likelihood' and 'significance' require a decision maker to have considerable foresight and judgement which may come with experience, but which also need to be assisted by a fuller listing of circumstances. For example, the Environment Protection Act (1974) of the Australian Commonwealth Government included the following:

a substantial impact on the ecosystems of an area;

a significant diminution of the aesthetic, recreational, scientific or other environmental quality, or value, of an area;

an adverse effect upon an area, or structure, that has an aesthetic, historical, scientific or social significance or other special value for the present or future generations;

the endangering, or further endangering, of any species of fauna or flora;

the curtailing of the range of beneficial uses of the environment;

the pollution of the environment;

environmental problems associated with the disposal of waste; or

increased demands on natural resources which are, or are likely to be, in short supply.

The last decade has seen a number of techniques for environmental impact assessment being proposed or used: some are described below. In general, however, they share a common approach which is often statutorily required. As a first step, the preparation of an inventory is necessary: a complete description of the environment as it exists in an area where a particular proposed action is being considered. However, the notion of 'completeness' has to be related to the character of the proposal, as also does the definition of the area being affected. Canter (1977) proposes two questions which set criteria for the inclusion or exclusion of a particular factor: (a) will the proposed action or any of its alternatives have an impact, either beneficial or detrimental, on the environmental factor?; and (b) will the environmental factor exert an influence on project construction, scheduling or subsequent operation? Not surprisingly, these criteria lead to different sets of factors being considered by different agencies since, in effect, they specify factors *after* deducing what the likely impact will be. Thus airports, nuclear power stations, gas pipelines, harbour dredging or motorway construction would all have different information requirements.

A description of the environmental setting of a proposed project needs a balance between comprehensiveness and selectivity. This, it can be asserted, is most likely to be achieved if the relevant phenomena are identified according to the expected (or possible) impacts of the project at various stages of its implementation. Thus some impacts may occur even as early as the planning and site-acquisition stages: for example, land uses might change through speculation in anticipation of the development. Though most are likely to be short-lived, some of the effects from these stages may persist. The following stages, construction and operation, are the ones most likely to leave permanent impacts, through the displacement or destruction of former conditions. Construction might affect drainage, or wildlife habitats. The operation of the project may generate noise, or air pollution. It might also have indirect effects on contiguous land uses, or create additional demand for housing and public facilities which, in turn, will have other environmental impacts.

The second step in an EIS consists of the prediction of the change in environmental conditions that will be brought about and an evaluation of their significance. To some extent, fairly accurate forecasts of physical impacts can be made using standard techniques and equations. For example, there is now considerable knowledge of the effects of water temperature on aquatic systems, of heavy metal concentrations and chemical defoliants on plants and human health, of construction works on geomorphic processes, and of large buildings and atmospheric pollution on micro-climates (Simmons 1974). However, in one sense this is 'general' knowledge which, in the context of an EIS, needs to be newly refined each time it is brought to bear on the unique set of conditions found at the location of a proposed development. Moreover, it would be a brave – or foolish – person who would claim that the knowledge of environmental systems is complete, even at a general level. At the time of writing a bitter debate is being conducted on the effects of 2,4,5-T (a chemical weedkiller and military weapon) on birth deformities. Often we only learn about environmental relationships through costly mistakes and, while we

should guard against them, it would be unrealistic to assume that the era of mistakes is over.

The other issue, evaluation of the significance of impacts, contains some ambivalence. From an ecological approach, 'significance' can be expressed in terms of impoverishment in the number and variety of species, or the creation of instability, or increasing hazard. Within this approach there are subtly distinct viewpoints such as the 'man as steward' theme and the 'duty to posterity' theme (Simmons 1974). The economic approach (Kneese 1977) is not wholly divorced from these – especially the latter – but differs because it is not so fixed on the very long term and also because it gives as much weight to the expected economic benefits which bring about the impact, as well as to the environmental costs arising from it. Neither approach fully convinces the adherents of the other, but the trend in environmental legislation has been to incorporate economic evaluation partly, one suspects, because it has an apparent rigour and because it lends some protection against ethical conservationist arguments.

The third step is the preparation of the EIS itself, in a format specified partly by law and partly by various guidelines. The US requirement is that the statement covers the following:

> the environmental impact of the proposed action;
> any adverse environmental effects that cannot be avoided should the proposal be implemented;
> alternatives to the proposed action;
> the relationship between local short-term uses of the human environment and the maintenance and enhancement of long-term productivity; and
> any irreversible and irretrievable commitments of resources that would be involved in the proposed action should it be implemented.

The proliferation of assessment techniques that has occurred in the last decade is partly a reflection of the variety of viewpoints and problems expressed in the previous paragraphs. While one may not always accept the tenor of Sewell and Little's (1973) assertion that 'the nature of impact staters would seem to deserve as rigorous analysis as impact statements themselves', it is still important to note that most techniques emphasise certain aspects at the expense of others. This is illustrated below with the help of three well-documented examples.

McHarg's (1969) overlay technique has already been referred to briefly (p. 39, above). Technically it is like the old sieve-map analysis applied by land-use planners (Forbes 1969), where the search for a suitable place to locate an activity was found by a filtering process (i.e. areas would be successively rejected because they contravened a series of criteria). In McHarg's approach the sequence of filters is collapsed into a single, composite evaluation but the effect is similar. Areas are graded according to their quality on each of a number of relevant factors (Fig. 8.4 and Table 8.4). By summing their scores on each factor, the area most suitable for the development is identified.

Although this is not strictly an impact-assessment technique (rather, it is an exploratory method for designating the use of land), it does employ the notion of impacts in its system of grading. McHarg differentiates, for example, between richest wildlife habitats and poorest, areas of greatest and least scenic beauty, and so on. By implication, the better the grading the more vulnerable an area is and the greater will be the impact of a particular development.

A number of problems are readily identifiable, such as what factors to include or exclude and whether or not scores or grades should be added together. But a more fundamental question hangs over McHarg's own viewpoint which is embodied in the technique. His preoccupation with ideas of limits, carrying capacities and 'non-negotiable' barriers to certain uses of natural areas lead him towards a form of environmental planning based on *absolute* advantage. The possibility that environmental degradation might be a cost worth incurring in order to secure an economic or social benefit is not entertained.

In this respect a contrast may be drawn with the matrix technique proposed by Leopold *et al.* (1971) and described in Canter (1977) which accepts the idea that there can be a trade-off between the advantages and disadvantages of alternative strategies (including a no-action strategy). However, having said that, it is unclear how such a trade-off is to be carried out.

The matrix technique contains two major features. The first is the matrix itself (Table 8.5), which is defined by the processes involved in a development (columns) and the environmental factors likely to be affected by the development (rows). In this example the processes in opening and operating a mine are shown: another development would probably have a different matrix. The second feature is the content of the cells in the matrix which comprise pairs of values describing the predicted *magnitude* and *importance* of each process on each factor.

Both features of the technique have raised problems. The matrix itself has been considered to be an inadequate representation where environmental impacts occur in a systematic, interrelated manner (Munn 1975). Presentation in the form of a matrix may lead to this complexity being overlooked. Then, within the matrix, there are issues over the definition of magnitude and importance. These employ a scaling (0 to 10) which is specific to each cell and there is no system of weighting which allows one factor to be compared with another. This leads to the further problem that the values cannot be manipulated (added, multiplied, etc.) in order to identify a preferred strategy. It is also unclear how 'importance' is to be defined. Although Spry (1975) has suggested it should refer to 'significance to the community', there needs to be further refinement to deal with short- and long-term impacts, local and regional impacts, and so on.

As a checklist to determine that relevant issues have been considered the matrix technique has some value. However, it only provides the roughest guidance on the comparative merits of different strategies. In this regard it is approximately equivalent to the planning balance sheet discussed earlier.

An attempt to go further and devise a method for making direct comparisons among strategies was undertaken by the Battelle Institute in the context of water-resource planning (Dee *et al.* 1973). Although this also uses a checklist, it does so in conjunction with a weighting system that estimates the contribution of each factor to overall environmental quality. Furthermore, it assesses the magnitude of impacts on a common basis, namely the proportional reduction in the quality of a factor caused by a particular development.

Application of the technique is illustrated in Fig. 8.5 and Table 8.6 which are derived from a pilot study carried out by Dee *et al.*, in which the impact of a proposed 315 ft dam and 32 mile reservoir on the Bear River in the western USA was assessed. Only two strategies were considered: building the dam, and not building it. The evaluation starts by identifying the factors considered to be relevant (78 in all) and assigning an importance rating (or parameter importance units – PIU) to each one such that the total rating sums to 1,000. This was done by quantifying the research team's 'subjective value judgements'.

At this point the dependence of the technique on a mixture of thorough scientific investigation and extreme subjectivity becomes most apparent. Each factor is recognised as having a spectrum of possible

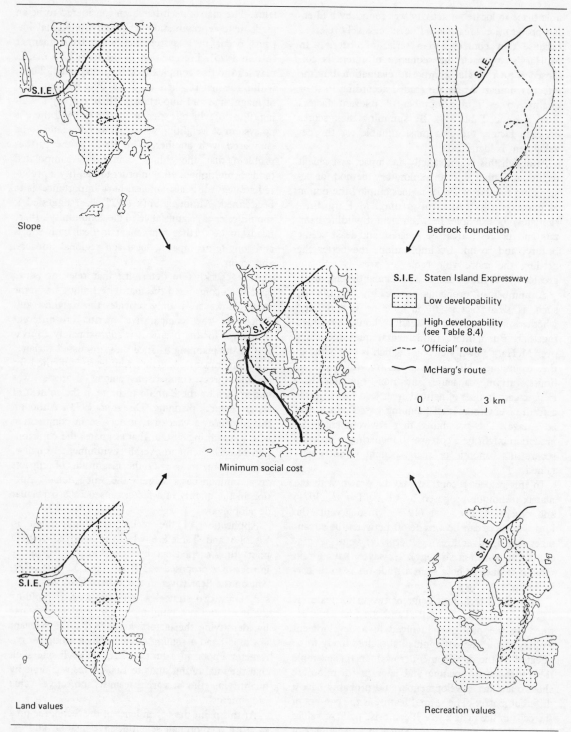

Slope

Bedrock foundation

S.I.E.   Staten Island Expressway

Low developability

High developability
(see Table 8.4)

'Official' route

McHarg's route

0   3 km

Minimum social cost

Land values

Recreation values

Fig. 8.4 The overlay technique for environmental impact assessment (McHarg 1969)

Table 8.4 Selected variables in McHarg's assessment of routes for the Richmond Parkway (after McHarg 1969)

| | least ← Developability → most | | |
Variable	Category 1	Category 2	Category 3
Slope	10%	2.5–10%	2.5%
Surface drainage	Surface water	Channels, constricted drainage	None
Soil drainage	Swamps, marshes	High water table	Good internal drainage
Bedrock foundation	Marshlands	Cretaceous sediments	Crystalline rocks
Land values	$3.50 per sq. ft.	$2.50–$3.50	$2.50
Recreational values	Public open space, institutions	Non-urbanised, high potential	Low potential

Table 8.5 A matrix analysis for environmental impact assessment

	Industrial sites and buildings	Highways and bridges	Transmission lines	Blasting and drilling	Surface excavation	Mineral processing	Trucking	Emplacement of tailings	Spills and leaks
Water quality					2/2	1/1		2/2	2/4
Atmospheric quality						2/3	1/1		1/2
Erosion		2/2			2/2			2/2	
Depositions, sedimentation		2/2			2/2			2/2	1/1
Shrubs					1/1				
Grasses					1/1				
Aquatic plants					2/2			2/3	1/4
Fish					2/2			2/2	1/4
Camping and hiking				1/1	2/4				
Scenic views and vistas	2/3	2/1	2/3		3/3			2/1	3/3
Wilderness qualities	4/4	4/4	2/3	1/2	3/3	2/5		3/5	3/5
Rare and unique species		2/5		5/10	2/4	5/10	5/10		
Health and safety							3/3		

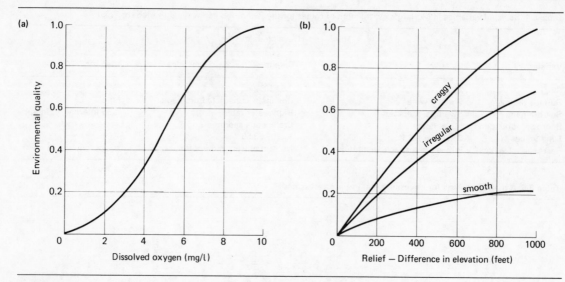

Fig. 8.5 Environmental quality; a parametric analysis of impacts (Dee *et al.* 1973)

conditions ranging from 'very good' to 'extremely poor'. The range is expressed numerically on a scale 1 to 0, and is related, where possible, to parameters which give rise to that variation. For example, water quality is considered to be affected by dissolved oxygen content in the manner shown in Fig. 8.5(a): a fall from 10 mg/ℓ to 9 mg/ℓ reduces quality of about 5 per cent. When social and cultural factors are subjected to the same system of scaling, the credibility of the technique begins to suffer greatly. This is unfortunate because it detracts from whatever objectivity is achieved in the analysis of physical environmental parameters.

Having defined these scales, it then remained to estimate what the situation would be before and after the construction of the dam. For some factors there was predicted to be no change; for others there would actually be an improvement as a result of the dam (e.g. crops, sport fishing, control of streamflow variation). But, for most there would be deterioration (see Table 8.6B).

Whether such deterioration is acceptable in order to gain the benefits from the dam cannot be resolved within the technique: this requires further analysis. What the technique can do, however, is indicate non-acceptability in instances where it can be predicted that an adverse impact will be great enough to reduce the quality of an environmental factor below an established standard. While this ought to be a feature of all planning control functions, it is not explicit in the other techniques discussed above.

In order to evaluate the different approaches to en-

vironmental impact assessment we must examine them in the light of two questions raised by all proposed developments: is a project worthwhile? and, which alternative course of development is preferred? On balance none of the techniques deals with the first question, but all of them make some progress towards the second, though they are not equally successful.

The value of a technique lies in its meeting several criteria (Warner and Preston 1974). All of them aim to be *comprehensive* in their selection of environmental factors. (Perhaps some go too far by including human factors that are impossible to define) and most of them try to use *explicit indicators* of environmental quality (the Leopold matrix is the weakest in this respect). All of them attempt to be *objective* and to be *replicable*, so that other researchers can test their findings or use their techniques in other studies. On the other hand, none of them distinguishes clearly between different *scales of impact* (local, regional, global, etc.); none gives adequate attention to *risks* and *uncertainty*; and none links impacts with different *social groups*. Lastly, they vary in their *flexibility* (Battelle was developed in the context of one environmental setting, while the others have a greater range of uses) and in their *resource requirements*. The last is a particularly important factor in the choice of assessment techniques where time, manpower and data are in short supply. Of the three examples discussed in this section, it would appear that the Battelle technique is the most demanding on resources.

Table 8.6 The Battelle system for environmental impact assessment (Dee *et al.* 1973)
A. **Elements in the system for the Bear River project**

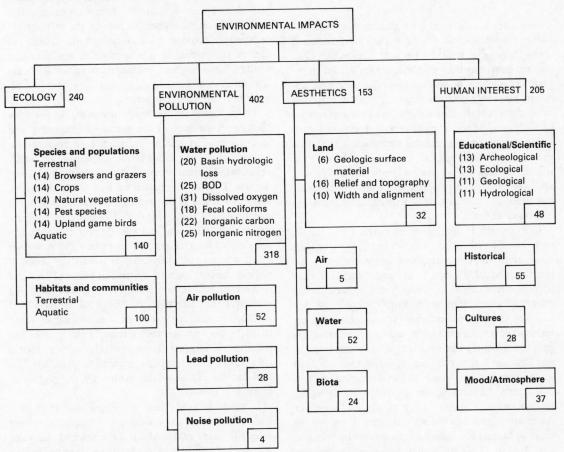

B. **Selected environmental changes**

Parameter	Weight (PIU)	Before project (EIU)	After project (EIU)	Change
Ecology				
Browsers and grazers	14	14.00	14.00	0.00
Crops	14	8.12	9.24	+1.12
Environmental pollution				
BOD	25	21.25	18.00	−3.25
Streamflow variation	28	8.68	19.32	+10.64
Aesthetics				
Geologic surface material	6	1.86	1.50	−0.36
Relief and topography	16	5.60	4.32	−1.28
Human interest				
Ecological	13	6.37	7.54	+1.17
Geological	11	3.52	2.75	−0.77

Geography and evaluation

Unless one is concerned with the most simple and narrow problems, all of the techniques discussed above require a multidisciplinary approach. They contain questions that only economists versed in investment analysis or welfare theory, or physical scientists knowledgeable about the qualities of materials or environmental relationships, could answer satisfactorily, and so on. They also pose questions which require at least a geographical viewpoint – and often the use of geographical methods. These arise whenever there are alternative locations for a project and whenever there is significant spatial variation in the incidence of costs and benefits, or advantages and disadvantages.

If we return to the information requirements for cost–benefit analysis listed earlier (p. 103) we can see three areas where geographers can contribute: the identification of impacts ('cost and benefit as they occur'), estimation of risks, and distributional effects. These information requirements are not peculiar to CBA alone: they are central to most forms of economic evaluation and environmental impact assessments. The fourth area, discount rates, is one that can happily be left to economists. So can some other questions such as the derivation of 'shadow prices' for non-market goods, or goods whose market price is distorted by tariffs, subsidies or under-employed resources. (Another item – the value of human life – has been discussed by economists (Mishan 1972) but would hardly rate as an area in which they have special expertise.)

This 'allocation' of roles should not be interpreted as a belief that geographers ought to confine themselves to subsidiary tasks of measuring and predicting tangible phenomena or that they should ignore values and qualities. It is, however, a statement of the obvious: that some disciplines are better equipped than others to handle certain questions. But such issues as willingness to pay, or the relationship between environmental quality and the quality of life, or the impact of a project on accessibility and welfare, have all been within the traditional collective interest of geography even if they have not always been approached formally.

Thus, many of the topics discussed in previous chapters contribute to the theme of evaluation and assessment. Firstly, the preparation of resource inventories is a basic input for describing the economic or physical environment within which an impact occurs. This is supplemented by the measurement and explanation of disparities in economic and social distributions, and natural and man-made environmental problems and hazards. Secondly, an understanding of interrelationships, systems and processes at all scales (and over time) is needed in order to project the incidence of impacts in terms of *how* and *where* they will occur, *what form they will take, and who* will be affected. A virtue of geographical methods in this context is that spatial analysis often helps to uncover indirect effects. Thirdly, the distribution of impacts, being spatially uneven, means that people are unequally favoured or penalised. To find out the extent of this requires a recognition of wants and ideals.

It would be quite false to suggest that evaluation and assessment techniques or geographical (and other) theories and methods have reached the stage at which unequivocal comparisons leading to the choice of worthwhile projects can be guaranteed. However, this theme, if followed, can provide a fruitful avenue through which to integrate many of the interests of geography.

9 Institutions, administration and policy

To western political scientists, institutions are devices for solving 'predicaments' (Finer 1970) through which 'demands' can be translated into policies (Blondel 1969), and for providing the machinery for 'rule making, rule application and rule adjudication' (Almond and Powell 1966). Each of these definitions shares a common feature, namely that it sees the need for institutions arising from the existence (or potential existence) of conflict over resources and activities among individuals and groups. This is not to say that institutions exist solely to resolve or suppress conflict: some act as channels for the expression of conflict. In serving these purposes they can vary widely in their effectiveness and acceptability. Hence it is not surprising that we should sometimes find as much conflict over institutions themselves as over the matters that they are supposed to help in resolving.

Geographers have tended to confine their attention to territorial aspects of institutions, though these can be defined quite broadly. Leaving aside geopolitics (described by Soja (1974) as perhaps the most painful 'burnt finger' in the discipline's history), their spectrum of interests extends over two major areas. The first is the territorial division of powers and responsibilities; the second concerns the use of territory. Some of the sub-headings into which these can be divided are listed as follows:

Horizontal separation of powers – regionalisation, local government boundaries, school and other service districts, electoral areas;
Vertical separation of powers – devolution and centralisation of executive and administrative roles;
Ownership, tenure and powers of acquisition over land; and
Regulation of externalities – land-use controls, environmental standards.

All of these items have been the subject of formal statute-making, particularly in more urbanised (and technologically advanced) societies. Indeed, it has been argued that urbanisation tends to engender conflict largely because anonymity makes reciprocity more difficult (Cox, Reynolds and Rokkan 1974). This may be an oversimplification (e.g. reciprocity may arise 'spontaneously' if the members of a community share a common attitude towards property values, as often happens in more affluent suburbs), but there can be little doubt that opportunities for inflicting nuisances or costs on others are related to population density.

The need for formal institutions, it has been asserted, arises when informal institutions prove inadequate in resolving certain kinds of conflicts (Cox 1974). In an informal, or de facto, system an optimal allocation of resources can be brought about by bargaining between individuals or groups. (The equilibrating market of perfect competition represents an abstract example of spontaneous conflict resolution.) However, where bargaining is inefficient, either because it is too expensive or because one party refuses to participate, then there may be justification for a formal, or de jure, arrangement to be imposed. Many instances can be cited, ranging from anti-pollution legislation in industrial countries to changes in land tenure (freehold replacing common ownership) in densely populated regions of developing countries.

As an explanation of the process of institutional evolution or change, Cox's argument is not entirely satisfactory for it ignores questions of motivation and control. That is, in addition to the breakdown or ineffectiveness of an informal system it is also necessary either that both parties should desire some new arrangement for settling conflict, or that one of them be strong enough to impose a new arrangement on the other, or that a third party should have power and authority to impose on both. In other words, a formal system is usually justified, at least implicitly, by net benefits to one party and preferably to all parties. Under this condition it is not difficult to predict that certain institutions, while working quite adequately, can be changed on the instigation of one party to reinforce its already strong position (Lukes 1974; Simmie 1974).

Moreover, with this in mind it is easy to understand how and why some informal institutions, while generally disapproved, remain unchanged. Examples that have received attention from geographers and sociologists are the informal division of territory among religious groups (Boal 1974) or racial groups (Rose 1970, 1978), or among rival gangs (Ley and Cybriwsky 1974). Even when the tenuous peace of these informal arrangements breaks down, as when riots or disorders occur (Adams 1972), it is not always obvious how a formal system could make things better.

Without in any way denying the importance and fascination of informal systems, either in their own right or as bases for comparison, it is on formal institutions that this chapter focuses. Also, as indicated earlier, most attention is given to matters that are directly concerned with territory. In the following sections the objective is to describe types of conflict that arise over the division and use of territory and to examine theoretical and methodological approaches to conflict resolution. Thereafter, the attention is shifted to the positive roles of institutions in creating opportunities for, and imposing constraints on, the formation of policy.

On conflict and geography

The common notion of a conflict is that of a 'trial of strength', not necessarily of force, between two or more *parties* (people or organisations) engaged in what Boulding (1962) describes as 'a situation in which the parties are aware of the incompatibility of future positions and in which each party wishes to occupy a position that is incompatible with the wishes of the other'. We can extend this to include conflicts over *principles*, where a party is faced not with an external antagonist but with a dilemma in which it has to choose among alternatives. In this case there may be no obvious trial of strength but all the other aspects of conflict remain the same.

Conflicts can be about many things: the use and distribution of resources, values and beliefs, relationships between individuals and groups, and so on. The ones that most interest us as geographers are those whose resolution directly affects (and may be affected by) the nature of human and physical landscapes. This excludes conflicts where the impact on landscapes would be the same whoever was successful (e.g. some personal contests over leadership) but it still leaves a great range of interesting situations in which the resolution of a conflict precludes a landscape developing along lines preferred by at least one of the proponents. Moreover, analysis of the nature and resolution of conflicts helps to explain why 'right' does not always triumph and, hence, why many social and environmental problems persist.

In addition to their variety of issues, conflicts take different forms and are resolved in different ways. A useful typology has been devised by Deutsch (1973), in which conflicts are classified on the basis of perception and experience (Table 9.1). Much of its value lies in its recognition that not every conflict is necessarily a *true* one. That is, the parties may misperceive its dimensions (*contingent* conflict), or focus on the wrong issues (*displaced*). The conflict may involve the wrong parties (*misattributed*) or it may combine more than one of these errors (*false* conflict).

Table 9.1 A typology of conflicts (Deutsch 1973)

Type	Objective conflict between A and B	Experienced* conflict between A and B	Type of misperception*		
			Contingency of conflict	Issues in conflict	Parties in conflict
I Veridical (true) conflict	Yes	Yes	No	No	No
II Contingent conflict	Yes	Yes	Yes	No	No
III Displaced conflict	Yes	Yes	No	Yes	No
IV Misattributed conflict	Yes	No	No	No	Yes
V Latent conflict	Yes	No	–	–	–
VI False conflict	No	Yes	Yes or Yes or Yes		

* The state of affairs as experienced and perceived by one of the parties to the conflict. The other party may, of course, experience and perceive it differently. Thus A may be displacing the conflict, and for B the conflict may be a latent one.

The other category is *latent* conflict, defined by Deutsch as 'one that should be occurring but is not' and which requires a raising of consciousness before being translated into true conflict. Here the challenge lies in capturing the essence of something that has yet to be given expression.

The distinction between true and contingent conflict is an interesting one because it depends largely on the dimensions of the action space of opposing parties. A narrow view might lead to competition for what appears to be a scarce resource in a small area, whereas a broader view might show that the resource is either not scarce or is better supplied elsewhere. Examples often occur in environmental debates: for instance, conservationists and logging companies may disagree about the use of a particular forest (Thier 1979). If the forest is the only source of timber and is a unique natural habitat, the conflict would indeed be true. On the other hand, if either the resource or the habitat is available elsewhere *and* if this can be brought into the reckoning, then it might be possible to satisfy both parties. Thus, contingency lies in the ability to change the premises of a conflict and, thereby, to assist in its resolution. We should not exaggerate the ease with which this can be done (often it is prevented by factors such as boundaries and organisation structures and local interests) but the identification of possible contingencies is an important task for geographers since many of them involve spatial shifts.

Likewise, it is important to recognise the occurrence of displaced and misattributed conflicts. The former might take the form of over-concern for symptoms rather than root causes. Thus, for example, redistributing children among schools, on the basis of colour or social class, may prove to be only a cosmetic treatment which diverts attention from the deeper issues of race relations or equality of opportunity (Leach 1974). Or it might be that a conflict is really about status and prestige, using a substantive issue as a platform. (Government departments appear to have a habit of indulging in this kind of covert warfare.) Misattributed conflict can be used as a diversionary tactic, to redirect attention from one target to another: usually it can be recognised quite easily, but it can prove difficult to combat. For example, packaging companies have developed the argument that they are innocent of pollution problems: after all it is not they who discard wrapping-paper, plastics and non-returnable glass bottles around the countryside and city streets (Dempsey 1974). They are quite right, of course; and it is difficult to sustain the counter-argument that the level of litter is a barometer of their profit.

All of these types of conflict are illustrated almost every day through the news media, though many of the most important cases actually receive little attention, either because they are very complex or because they are not very newsworthy. An example is the issue of geographical fragmentation of metropolitan government. This is a debate that has occurred (and still occurs) in many countries (Hicks 1974) but which has been a major preoccupation in the United States (Williams 1971; Cox 1973, 1978). It also applies to non-urban areas, but in more muted form.

The problem of urban fragmentation is often perceived as that of a conflict between potential benefits to be gained from economies of scale on one hand, and the desire for local democracy on the other. That, of course, is an oversimplification, for as Cox elucidates in the case of American cities, fragmentation is associated with the entrenchments of inequalities, particularly between central cities and peripheral suburbs. Suburban local authorities often have the power to determine their broad social and income structure through 'exclusionary' zoning in the form of minimum lot sizes, single-family residences, building codes and rating systems. These are formal institutions, but they can be reinforced by informal ones such as covertly discriminatory housing markets. The ensuing inequalities tend to take the form of disparities in the ratio of local tax resources to local expenditure burden, and consequently, in disparities in the quality of services available to inner-city and outer-suburban residents.

Given this situation (which is not confined to the USA), it is rational to ask whether improvements in both efficiency and equality could not be brought about by the consolidation of local government, possibly into metropolitan governments of the kind found in Toronto or London, and a redistribution of responsibilities and powers. In this way it might be possible to reduce, if not to eliminate, the disparities between resources and burdens found in different areas, and to remove discriminatory zoning. It might also create conditions for scale economies in the provision of public facilities. This is broadly the major remedy to fragmentation to be found in Cox's conclusions and, in concept at least, it is a plausible solution.

However, if fragmentation is responsible for one set of conflicts then its proposed remedy – consolidation – brings about another set. This would be the case even if the net social benefits derived from the potential scale economies of consolidation were sub-stantial and certain. Firstly, it is always to be ex-pected that fragmentation and consolidation work to the economic advantage of different sections of the population. It is unlikely that richer areas would wil-lingly give up their real privileges, even for the great-er good, if this meant a reduction of their resources or an increase in their taxes. Secondly, consolidation also means changing relationships and patterns of control, with formerly relatively independent sub-urbs becoming part of a larger government machine.

The purpose of the last few paragraphs has not been to enter an argument about the relative merits of different institutional systems. Rather it has been to point out that any single system is likely to arouse conflicts and that any proposal to change a system is likely to raise new conflicts. These are inescapable.

The way in which conflicts can be analysed de-pends on the degree to which participants and their objectives can be identified. In many issues this can-not be done sufficiently accurately to allow more than a verbal description and prognosis, sup-plemented by whatever empirical evidence is avail-able. This is the case in the example discussed above – metropolitan fragmentation. Other situations occur, however, where a quantitative analysis is worth-while. The results will not be binding on the participants but they allow third parties to gain some insight into the nature and possible course of a true conflict.

In Chapter 3 an example was described of quan-titative conflict analysis where the parties and their wants were identifiable. This example, which can be used again, dealt with attitudes towards a highway, subway and bridge project in Washington, D.C. (Saaty 1978). The preferences of the parties were identified (Table 3.2 above), allowing the observer to judge how the parties would fare from a variety of outcomes to the conflict, and to predict how each party would react. The crucial point to seek is whether the conflict can have a stable outcome. Sta-bility means that a party cannot improve its position (e.g. by bringing sanctions to bear on the others) since other parties can respond with equal or greater sanctions. Moreover, stability is consistent with an impasse (or perpetual conflict) in which the original position, despite being satisfactory to no one, cannot be improved upon. Finally, the analysis requires that the preferences of parties are fixed.

In addition to an identification of preferences, Saa-ty took account of sanctions that each party might bring to bear in the case of unacceptable outcomes.

Thus, public officials could veto some parts of the project being undertaken without others, conserva-tionists could litigate, and inner-city residents (being poor and black) could only resort to civil distur-bances.

Using this information, plus that in Table 3.2, Saaty conducted analyses using a modified form of game theory (see p. 76, above) in which different outcomes were tested for their stability. The results were illuminating. If the groups acted individually there would be an impasse; any other outcome would be unstable. One way out would be a coalition be-tween the public officials and *either* the inner-city re-sidents *or* the conservationists, although in each case the end result would still not be the full package of proposals and there would still be some danger of sanctions from the party excluded from the coalition. In reality no coalition was formed, but the public officials came closer to compromise by deleting some of the controversial items. This resulted in reprisals from the federal government which withheld build-ing funds, and thus contributed to another – this time unforeseen – impasse.

Spatial administration

It is never possible to design, or redesign, institu-tions and territories that are wholly acceptable to everyone. When it comes to the substance of political control and the distribution of resources, the out-come has less to do with cold logic than with per-ceived self-interest. Whatever the motives, the effects of allocating resources or of drawing bound-aries for the allocation of power cannot be entirely neutral: costs or benefits (or both) fall unequally upon communities and their members.

Yet there are times when new systems have to be introduced or old ones reformed. Pressure to do so may come from a wide range of sources. It may be built into the system in the form of a constitutional or statutory requirement to review boundaries, as is often the case with electoral districts. Or, secondly, the pressure may come from interest groups such as developers or manufacturers lobbying for a relaxa-tion of environmental standards. Or, thirdly, it may come from less overtly partisan groups seeking changes whose benefits are more diffuse but none the less biased in one direction or another: local govern-ment reform societies and environmentalist movements belong to this category. On occasions the pressure may be for radical change: at other times it may be for marginal adjustment. In both instances there is a role for an applied geographer, either as direct participant or as a disinterested observer and reporter (Perry 1969).

At the outset, however, it needs to be observed that most geographical work on spatial administration has been directed at its more simple aspects: for example, defining single-function areas based on lim-mited criteria. Also, it has attemped to maintain a neutral position. These are not without value, for simplicity and neutrality have enabled some useful experimentation with conceptual and methodological approaches to what, otherwise, are difficult prob-lems. Where the difficulty arises – and remains to be overcome – is in contributing meaningfully to more complex issues related to multifunctional areas and hierarchical administrative (and political) rela-tionships (Küpper 1973). In the following para-graphs we shall follow the transition from a simple to moderately complex situations, discussing concepts and methods that have been applied.

Despite their importance for the working of demo-cratic systems, the definition of electoral areas or constituencies is conceptually the most simple of all spatial divisions. Using the principle of one man/one vote/one value, the task is confined to dividing a territory into a statutorily determined number of constituencies such that each contains a similar num-ber of electors. However, the simplicity of the task is more apparent than real, for there are many ways in which the same objective can be reached, and elec-toral division and redistricting can be accompanied by severe constraints. It might be decided, for exam-ple, that areas should be compact, thus minimising the distances between electors and (say) the central points of electoral areas. This might give electors easier access to their representatives. Or, this con-straint might be accompanied by another which re-quires that all areas contain a balance of interest or – the opposite case – that they be homogeneous in social and economic structure. These constraints can be handled with a fair amount of objectivity, but this is less true of another – that redistricting observe the partisan demands of political parties, most of whom, given the chance, would show no hesitation in gerry-mandering an electorate to their own advantage. Thus electoral redistricting may also have to take account of the geography of voting as well as ques-tions of representation.

Basic approaches to redistricting have been de-

scribed by Morrill (1973, 1976) and Taylor (1978). In Morrill's case there is added interest for he deals with his own involvement in redrawing electoral boundaries in the state of Washington, which arose after the two political parties had each prepared a scheme that was unacceptable to the other. Morrill was required to work within the following con-straints:

to allow a maximum of only 1 per cent (plus or minus 685 around 68,445) deviation in population per electoral area;

to maintain to the greatest extent possible the in-tegrity of countries, cities, census county divi-sions, and census tracts. Enumeration districts outside urban areas, and blocks in urban areas were the smallest units permissible. Electoral pre-cincts were excluded as possible units, since no census data were collected for them;

to form the districts as compactly as possible and to avoid unnecessary irregularities or sinuosities;

to avoid crossing natural geographical barriers, especially such bodies of water as Lake Washing-ton and Puget Sound; to extend no district across the crest of the Cascade Mountains except along the Columbia River; and,

to reflect, as far as possible, unity of character or interest; in particular, Indian reservations should not be split, and as much of the Seattle 'Central Area' (black ghetto) should be included within one district as possible.

Morrill was faced by a choice of techniques. One might have been to work from existing boundaries and merely shift 'surpluses' from overpopulated electorates to underpopulated electorates until a bal-ance was achieved. This was rejected, mainly be-cause population changes had been so great since the previous redistricting that the imbalances were too large to be handled in this way. Hence a method was chosen which aims at the creation of an entirely new set of constituencies. Using an iterative programme, the method starts with a set of centres (e.g. major cities), equal to the number of constituencies, and assigns census tracts to these centres using a mini-mum-distance criterion until the entire population has been allocated. The areas so derived are then ex-amined to find their median centre (i.e. their point of minimum aggregate travel), which may not be the original centre. These new centres then replace the first set and the population is reassigned to them, using the same distance-minimising criterion. The procedure continues until a new set of median cen-

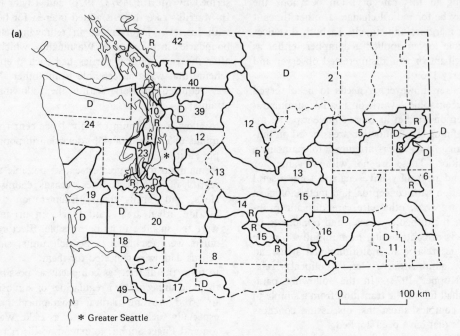

* Greater Seattle

Democratic plan for redistricting Washington state

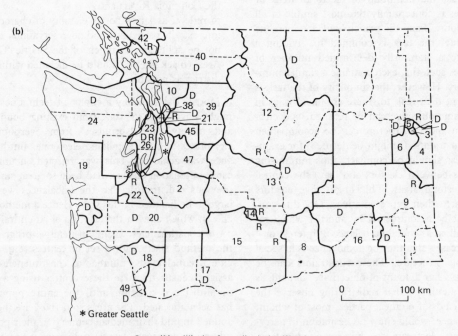

* Greater Seattle

Court (Morrill) plan for redistricting Washington state

Fig. 9.1 Political redistricting (after Morrill 1976)

tres is approximately the same as the previous set. Morrill's solution (Fig. 9.1) shows considerable departures from both the Democratic and Republican proposals but, using various indices of compactness, he was able to show its superiority on geometric criteria. It also turned out to be a non-partisan solution as it did not significantly concentrate the support of one party any more than the other.

It may, at first, appear strange to assert that political redistricting is conceptually more simple than administrative redistricting, where (say) school catchment areas or electricity supply areas are involved. However, the latter possess one feature not found in the former, namely that the number of such areas (and of supply points) is not statutorily fixed. Faced with this, the administrative decision requires not only that it serves social cost objectives, such as maximising the accessibility of the users to the facility, but also that it serves operating cost objectives, where these costs are known to be related to the size and structure of the facility and its district. A further difference lies in the degree to which administrative decision making is incremental. It is unusual to find an entire scheme for any service being created *de novo* or undergoing total reform. More often it is a case of a new facility being created or an old one removed, or of marginal adjustments in the size of existing ones (Teitz 1968). Not surprisingly, therefore, we find entrenched patterns of inequality in some services. For example, the distribution of hospital beds in many large cities reflects the propensity to invest in existing hospitals, giving the older suburbs a much higher bed–population ratio than the new suburbs.

Questions of scale and sequential investment are both crucial in spatial administration. Unfortunately, both have tended to be neglected in the pursuit of optimising techniques in the form of location-allocation models based on relatively simple, static conditions (Scott 1971; Massam 1975). This is not to detract from the importance of these models but only to stress the range of information required in making the fullest use of them. In addition to the distribution of population and the long-run cost schedules of services, they also need to take account of the disutility that people put on travel and distance from facilities. Even with this information, location-allocation models may do no more than show only one optimal arrangement of facilities at one point in time.

The issue of sequential investment may be illustrated by Scott's preference for what he calls a 'backward recursive approach' (rather than a 'for-

ward' one) to the location of facilities. Given an objective (say, to locate three fire stations at a rate of one per year), he argues effectively that it is more efficient, in terms of the operation of the entire system, to find the best set of locations simultaneously and then to find the best sequence for developing them rather than to find the best single location, followed by the next best (for the second location) and then the third. As a geometric solution to a static hypothetical problem this is unexceptionable. But it is only workable within the limits of the situation, whose conditions include certainty in the rate of investment in fire stations and in the future distribution of demand for fire services. If certainty is lacking – as can happen in the best of budgetary systems – then one can easily imagine the temptation to adopt a forward recursive approach.

In dealing with other issues – size of administrative areas – it is tempting to regard as optimal that area which has the best combination of 'jointness efficiency' and 'distribution efficiency' (Cox 1974). The first of these contains two components: scale economies and externalities. An ideal area would thus be large enough for long-run cost curves of public services to be minimised and for external economies to be confined to residents of the area. For distribution efficiency to be maximised it is required that actual consumption of a public good is equal to effective demand (i.e. marginal private costs equal marginal private benefit). Thus, *ceteris paribus*, an area which is too small would lose jointness efficiencies while one that was too large would have unequal distribution of services with some zones being poorly supplied.

This concept of a trade-off between types of efficiency is attractive and has indeed formed the basis of arguments for local government reform and 'metropolitanisation' (Barnett 1974). However, there are problems which hinder its direct applicability. Firstly, it is not always clear that a trade-off actually can take place even in a single activity. Some activities may not have clear scale economies (Massam 1975). If this is the case, the optimum for the particular function is indeterminable. Secondly, even if optima could be defined for every function, each one would be different. Since local government deals with many functions simultaneously it would be inevitable that some, if not all, would be operating on a sub-optimal scale. Thirdly, the questions arise of who participates in a trade-off? and what are the objectives of participants? Even if the answer to the first was 'everybody', the situation would not be clear cut, because each individual has more than one objective. For example, as ratepayers we might seek jointness efficiency and as consumers we might prefer distribution efficiency – provided we are not subsidising others. But the answer is unlikely to be 'everybody' – or at least everybody equally. Most reforms are motivated and directed by those most directly concerned with managing the system (Pahl 1970) or those who stand to gain most. Their weighting of objectives is likely to be different from that of the rest of the community.

Although jointness and distribution efficiencies are conceptually important in the evaluation of reforms it is not always obvious that these are taken into account when reforms are contemplated. In Sydney, where there have been referenda to determine whether neighbouring areas should amalgamate, the negative vote has tended to be the majority in both richer and poorer areas. Motives are hard to determine, but the vote suggests that jointness efficiency does not carry much weight with electorates. In this case, however, one cannot confidently suggest distribution efficiency was the prime factor since many functions are administered by the state government, and other factors making for localism might have dominated. Not surprisingly, therefore, where large-scale reforms have taken place, as in England and Wales, they have been imposed on local government and to a considerable extent have been based on bureaucratic criteria with political overtones. The system eventually chosen gave rural areas a little more influence than the proposals that were rejected (Glasson 1974; Broadbent 1977). Had there not been a national election in 1974 which the urban-oriented Labour Party lost, the system would probably have been less biased in this direction.

The bureaucratic significance of local government reform is illustrated in Fig. 9.2, derived from Bennett and Chorley (1978), which shows the association between number of decision units, types of linkages and numbers of interrelationships between units. It says nothing about costs and benefits. Reforms directed towards amalgamation tend to do two things: they reduce the number of decision units and they change the linkages between them. To facilitate the latter requires a redistribution of functions and powers and, in so doing, may cause a reduction of the ability of lower-tier authorities to bargain with each other (type 1 in Fig. 9.2). Or it may lead to retention of lower-order functions by a reduced number of lower-order areas – or a devolution of

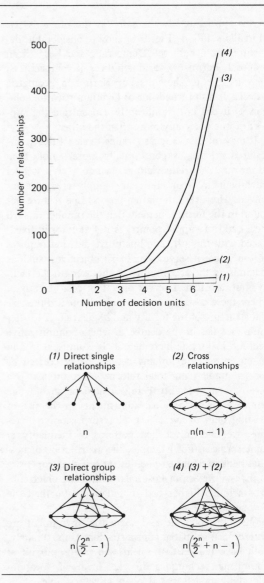

Fig. 9.2 Fragmentation, organisation and complexity in government systems (Bennett and Chorley 1978)

these functions – with no influence from the higher level (type 2).

Spatial administration is also concerned with practical aspects of regionalism and regionalisation. Although these involve some of the questions discussed under the organisation of local government, the particular significance of regions is that they form an extra tier between central and local levels. Their roles can vary widely from being a convenient means of collating statistics, to providing a geographical basis for central government planning, and, ultimately, to serving a political function representing local interests. Their apogee is found in political federalism, where rights and responsibilities are allocated exclusively to central and state governments (Wheare 1963). Federation has proved to be an attractive concept – especially in non-federal countries – and, in recent years, has been accompanied by devolutionist movements. In Britain these were taken seriously enough to warrant a Royal Commission on the constitution. However, the arguments of the devolutionists have not persuaded everyone. For example, opposing the case for devolution of power to a regional level in Britain, Donnison (1974) argued not only that regions were ill-defined but also that it was questionable whether there were any functions that could not be handled at either the local or the central level: in other words, regions would be unnecessary encumbrances. One might go further and ask whether regionalisation may not divert attention from areas which share common problems but are spatially dispersed.

Political devolution is only one possible purpose of regionalisation, and it tends to be less observed than discussed. What is more common is the action of central governments to compartmentalise functions on a spatial basis for administrative, budgetary, informational or operational purposes. Occasionally, also, one finds local governments combining on an *ad hoc* basis to give greater force to their representations to central government; for example, the Standing Conference on London and South East Regional Planning or the Western Sydney Regional Organisation of Councils. These tend to be formed out of a loose coalition of interests and although their impact may be considerable, their membership tends to be voluntary and their status is that of a lobby group concerned with a narrow range of issues. The Standing Conference was primarily concerned with physical planning and the Western Sydney Organisation with employment and social infrastructure.

Embarking on administrative regionalisation without clearly stated aims and constraints makes it difficult to evaluate the outcome in realistic terms. This is not a task for which the more familiar classification and regionalisation procedures (Grigg 1965; Johnston 1968; Abler, Adams and Gould 1972) are sufficient by themselves. Firstly, the principles of classification – minimising variance, being spatially exhaustive and avoiding ambiguites – are not always required. For example, environmental legislation in

New South Wales (NSW) allows the minister to declare regions for the preparation of plans: these are *ad hoc* regions, which may overlap with ones used in previous studies. Secondly, even when they are required, these principles need to be used in conjunction with the aims of the proponents of regionalism, the constraints of political, environmental and economic systems, and a clear understanding of the character and interrelationships of the functions to be reorganised. The danger of proceeding without these is shown by another Australian case (Logan *et al.* 1975). Using factor analysis and grouping algorithms, the country's 900 local government areas were reduced to a much smaller number of regions which nicely summarised the spatial structure of the Australian economy. What the report lacked was an evaluation of the appropriateness of these regions for the organisation of governmental functions. This study is far from being an isolated example (e.g. Brunn 1974).

Some of the issues and problems of regionalisation can be illustrated by a practical example in which the author had a minor role. This concerned the reorganisation of NSW State Government administration, of which the question of regionalisation was but one part (Wilenski 1978). The functions of the state are distributed among about 70 departments and instrumentalities. Most are highly centralised, in terms of both employment and control, in Sydney. Of those that have regionalised, following earlier initiatives from state government, few have common boundaries throughout the state or common structures in terms of responsibilities and powers devolved upon regional administrators: in one department any payment greater than $10 had to be authorised by central office while in another there were wide-reaching powers over expenditure and employment. Because the state is so large, the degree of centralisation creates the further problem of poor accessibility of many areas to the place where decisions are made. However, it was made clear that regionalisation only meant organisational change and that significant decentralisation of government employment or devolution of responsibilities were not to be part of the enquiry.

There were two main arguments in favour of regionalisation. The pre-eminent one was that advantages lay in more efficient information gathering, budgeting and resource allocation within departments and the greater ease and flexibility of cooperation among departments. This advantage was

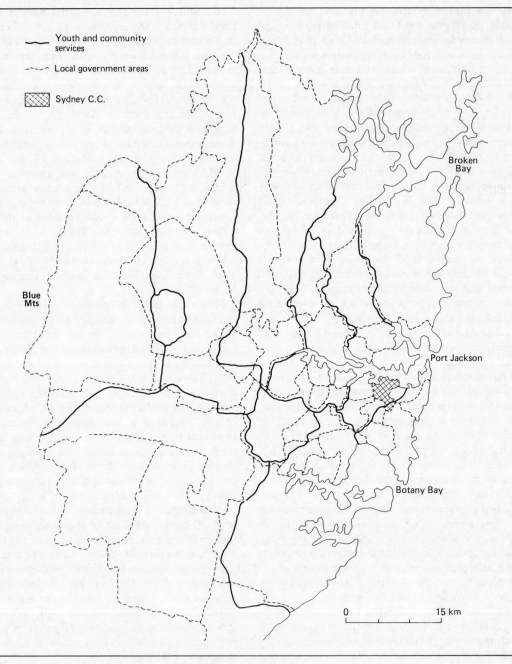

Fig. 9.3 Nonconformity of boundaries in government services: Sydney, 1974

promoted not only by some senior departmental representatives, but also by representatives of less well endowed regions who anticipated a greater allocation of resources following from a regionally organised central administration. Secondly, a case was also presented that it would allow increased effectiveness in the delivery of services to the community and more public participation.

Assuming both arguments to be valid, the questions arise: what form of regionalisation best serves them? Should all departments be regionalised, or just some? Should they all use common boundaries? What criteria should be followed for defining regions? None of these were conclusively answered but some general features emerged. It is clear, firstly, that the two arguments do not pose the same demands on regionalisation. Although both would be best served by conformity of boundaries the first tends to favour large regions without much further subdivision whereas the second favours much greater localism. Secondly, not all departments are amenable to regionalisation. Some had already been exempted by Cabinet decision (e.g. the Central Mapping Authority, The Division of Cultural Activities and The Registrar General's Department), and some found the existing set of standard statistical regions inappropriate (e.g. the water and electricity authorities). Thirdly, a problem arose over division of the metropolitan region which, though smallest in area, contains three-fifths of the state's population. For the purpose of regional planning this is best left as a single unit, but the proponents of the service delivery argument saw a need for division. Unfortunately, the region lacks a ready-made set of sub-regions and ministries tended to work to unique boundaries (Sutton 1974); see, for example, Fig. 9.3.

Whether the NSW government will ultimately adopt a regionalisation policy in line with any of the cases put by its proponents remains an open question. However, one is prompted to ask whether the regionalisation debate is not a 'red herring' and that what ought to be at issue are the functions and spatial organisation of local government.

Levers and constraints

Regulative and indicative planning – the two most common types in western countries – do not only restrict the freedom of individuals. They also involve the creation of rules and procedures for the public sector, and a definition of the limits of its authority. Public actions cannot take place in a vacuum. At the very least they need to be accommodated within a consistent framework of institutions. This may require new ones to be created or old ones to be amended or abolished: or, more likely, it will mean fitting into existing institutions. In each case there is a task for geographers, either in assisting in the formulation of institutions or in interpreting the opportunities (or levers) and constraints which they contain.

Although levers and constraints are equally important, we would normally expect that it would be an easier task to draw up a list of levers. These would not only be specific to each policy or plan but would also involve positive actions whose parameters can be (and need to be) identified. In general terms, policy levers comprise the following:

direct controls on private and public agencies and individuals;
incentives and promotional activities aimed at influencing behaviour in certain directions; and
manipulation of public resources in pursuit of given strategies.

The identification of constraints poses a different problem because, in many cases, they require foresight to predict the responses which plans or policies might evoke. At this point it is possible only to generalise, and to indicate three broad types of constraint which would be likely to face any planner or policy maker. These are:

those arising from informal behaviour and beliefs, tastes and desires, and embedded in existing settlement and interaction patterns;
those embodied in patterns of, and attitudes towards, control and ownership of resources; and
those enacted in existing legislation.

To illustrate how levers and constraints are perceived it is useful to concentrate on a set of studies which focus on a related set of planning problems within a single region. Three such studies, referring either to the whole or to parts of East Anglia, were conducted within a short space of time in the 1970s. One is a central government study, one is a county council study, and the last is a piece of academic research commissioned by a central government department. In substance they range from a strategic policy, through a settlement plan, to a study of the problems of rural accessibility.

The region as a whole had had, for over a decade, one of the fastest rates of population growth in Britain, and this was thought likely to continue, as a result of regional migration, particularly to planned growth centres. At the same time, it showed uneven development, with its southern and western parts growing fastest and most rural areas generally losing to the cities in the region (Sant and Moseley 1977). Hence the importance of the three studies to the future of the region.

Sub-regional reallocation

The East Anglian Strategic Plan (Department of the
Environment 1974) was prepared in order to set out
objectives and to identify broad economic, social and
land-use policies which could guide development. To
do this the planning team first made a projection of
likely population trends (Fig. 9.4) and examined a
variety of regional issues and problems. It then defined a set of alternative patterns of development
each of which diverged from the most likely trend
(Fig. 9.5). The aim was to inquire whether any of
them might better serve the various objectives and to
identify a 'preferred' policy. It can be seen that each
of the alternatives involves a different distribution of
population and, therefore, of resources. These were
subjected to a qualitative evaluation using a modified
form of goals-achievement matrix (see p. 110), from
which a preferred strategy emerged which combined
elements of strategies B and D. That is, it was proposed that development be diverted so that it would

be greater than predicted in the north and east of the
region (except in the vicinity of its major city, Norwich) and less than predicted in the south and west.
It was assumed that diversion would not affect the
total level of growth, though this is debatable since
so much is dependent on external migration.

Now, whatever the merits of such a strategy the
crucial questions are whether and how it could be
carried out. Direct controls on migration and the
natural increase of population are not part of the
armoury of policy makers, so anything to effect a
change from the predicted distributions of population would have to be done indirectly. Moreover, it
would have to be done with the support (or at least
the acquiescence) of a number of public and private
bodies which were either controlled or had their
main connections outside the region. Some of these
were identified, perhaps a little cynically, by the
director of the strategic planning team: 'the Regional
Water Authority with its involvement in sewerage investment and hence in the location of new housing;

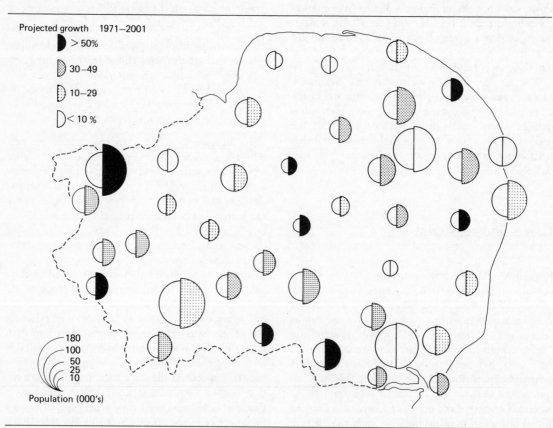

Fig. 9.4 The distribution and growth of population in East Anglia: a forecast for 2001 (Sant and Moseley 1977)

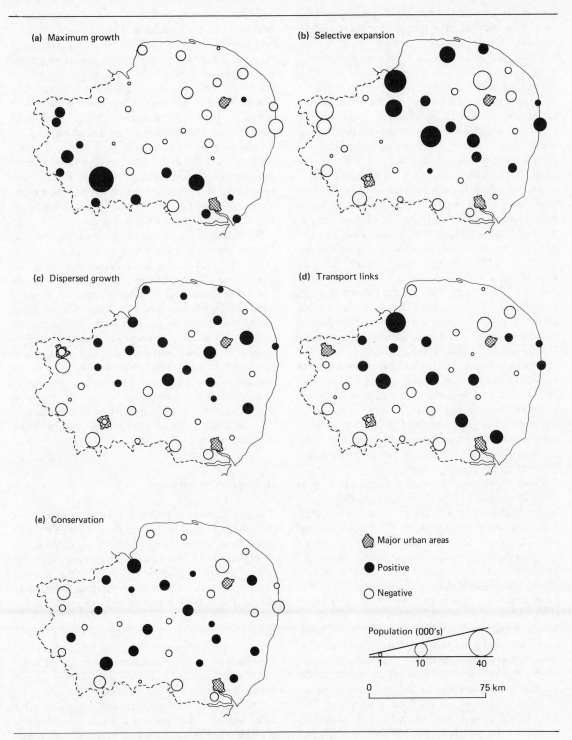

(a) Maximum growth

(b) Selective expansion

(c) Dispersed growth

(d) Transport links

(e) Conservation

Major urban areas

Positive

Negative

Population (000's)

1   10   40

0   75 km

Fig. 9.5 Alternative strategies for regional development. The maps indicate positive or negative departures from the 'control pattern' of growth depicted in Fig. 9.4. (East Anglia Regional Strategy Team, 1974)

the Greater London Council, which controls in large part the movement of industry, population and investment into the region; the Ministry of Agriculture, with its power to influence the scale and nature of agricultural activity in the region; and the Felixstowe Dock and Railway company, whose expansion in large part determined the pattern to trunk road investment' (Sant and Moseley 1977). To this list could be added local government authorities, responsible for a large part of public expenditure and for planning controls, and private individuals who often have a perverse disregard for strategy plans.

A plan which diverts population from its most likely distribution needs to be accompanied by an appropriate diversion of public investment, in services and infrastructure (which, *inter alia*, provides jobs) and private investment, which also creates employment. (The planning team argued that the volume and distribution of private investment would continue to be closely related to the pattern of public expenditure.) Thus one needs to know what margin of public expenditure is available for discretionary redistribution, and how feasible such redistribution is. For example, while there might in principle be freedom to locate a regional hospital anywhere in the region, in practice the number of appropriate locations is likely to be narrowly limited to a few major centres – the very places which a preferred strategy might wish to avoid. Associated with the question of discretion over the distribution of expenditure is the exercise of control over physical planning as a means of directing the private sector.

Flows of expenditure in the public sector pass through central and local channels. Although the *quality* of public services is, to a large extent, determined by current expenditure, the crucial element affecting long-term distributions of population and activities is capital-account expenditure. Room for manoeuvre exists in both cases, but on a limited scale. The major spending departments of central government are constrained by budgetary allocations and have to operate within their ranking of priorities and needs across the entire national economy. Any significant and continuing bias towards a particular sub-region needs to be justified against the claims of other sub-regions, as well as against the department's own operating criteria for the efficient use of public resources. At the local government level most capital projects are financed out of borrowing, permission for which is required from central government. They are therefore subject to two constraints; the borrowing capacity of local authorities

and the monetary policy of central government, which can be quite unstable. Local government current expenditure is mainly derived from rates (local taxes) and the rate support grant paid from central government resources. The first of these tends to favour the areas that are already faster growing, thus reinforcing their apparent potential for growth. The second is fairly neutral, though it can be used to promote, in a mild way, a regional strategy. This occurs because, although the formula for the rate support grant is based mainly on population characteristics (the 'needs' element) and the disparity between local rate resources per capital and the national average (the 'resources' element), it can be altered, though to do so is not an easy task.

What emerges from this discussion is *not* that: (a) because resources are difficult to divert or redistribute, (b) therefore the preparation of regional strategies is a waste of time. The former is true but the latter does not necessarily follow. Marginal adjustments, sustained over a period of a decade or more, may cause a redistribution in the desired direction. In addition, a major investment in a pivotal sector (e.g. a new transport link) may initiate more rapid growth in an area. However, the point to be stressed is that a regional strategy has to come to terms with two fundamental issues: the operating criteria of the agencies involved in resource allocation, and the political environment within which reallocation must take place.

Hierarchical selectivity

The second example provides a contrast in the levers that can be brought to bear in the pursuit of a policy when the proposing body also has close control over development. In this case, the *Structure Plan* (and preceding *Interim Settlement Plan*) of the Norfolk County Council (1977) was prepared in response to the need to co-ordinate the location of investment in housing, industry, services and utilities. These are much the same items as the Strategy Plan discussed above, but it had to descend to the detailed level of individual parishes and to discriminate among small communities on the basis of the scale of development believed to be appropriate for them.

The county's rural problem was argued to have three aspects. The first was a decline in male employment and outmigration of young people consequent upon continuing changes in agriculture. The second was a conflict between, on one hand, the economic advantages of concentrating population in fewer places and, on the other hand, the pressure for dispersal of new residential development (mainly attracted by cheaper land) into a large number of small hamlets and villages where the costs of providing services are higher. Much of this development was associated with retirement migration which could be predicted to create further problems as elderly people become less mobile and, perhaps, more dependent on support services (Sant 1977). Thirdly, the threshold population needed to support facilities in rural areas was asserted to be increasing as a result of economic and technical changes, and of changes in policy. It was argued that these affected such important activities as education, health services, public transport, utilities and retailing.

To meet these three trends it was proposed that the structure of rural settlement should be controlled, by stimulating and concentrating development in certain locations – selected towns and larger villages – accessible to more truly rural areas in which development would be quite severely restricted. In essence, the plan was described as an attempt to bring a 'medieval central place system into the late twentieth century'.

The pattern of public goods is a crucial element in this plan. Although a basic level of expenditure is predetermined by the existing settlement pattern, local governments also possess some powerful levers which they can implement within their own territories. The direction of capital and, to a lesser extent, current expenditure is one lever. Concentration of certain activities can have significant effect on residential preferences: for example proximity to modern primary schools is important to young families, and availability of subsidised public transport is valued by those without private cars. An equally important lever is direct control over development through the use of planning permits. Thus where a village falls in the 'wrong' category a council can resist development in excess of service provision by rejecting applications for the construction of houses, whether they be individual dwellings or substantial estates.

Given these powers, the crucial question is which villages or areas to stimulate and which to restrict. The policy that was ultimately accepted was based on the location and hierarchical status of villages. In general, the larger the place the more suitable it was thought to be for development. At the same time the larger places located in the region's problem areas were given priority (Norfolk County Council 1977).

Personal accessibility

The third example continues the theme of rural planning and focuses on problems of *personal* mobility and accessibility. Against the background of changing distributions of people, jobs and services, and of declining public transport facilities, a study was commissioned to consider actions and policies that might be taken by various levels and agencies of government (Moseley *et al.* 1976; Moseley 1979).

Having analysed the trends and the spatial and social variations in mobility and accessibility, the study concluded that there should be a shift in focus away from *contextual* elements of the problem (e.g. the decline of bus services or low car availability among the elderly) towards a statement of *problematic* social group/activity combinations (e.g. elderly/post-office, teenager/urban entertainment). Such an approach, it was believed, would provide a sounder basis for the derivation, evaluation and monitoring of policies. After further analysis of the problems, and before recommending policies, the study gave considerable attention to the options that were available and the levers and constraints affecting their implementation. Two 'themes' were readily apparent. One was a division of options between those that are transport-based and those that are not. The other was a distinction between options that might be applicable and effective in the short term and those that require continuous management and resource allocation over the long term. Broadly, the options were listed as:

measures designed to influence rural population distribution, to promote concentration in, or along routes to, settlements which are large enough to support a range of private and public services (this corresponds roughly with the settlement policy discussed above);

actions which rearrange the spatial and temporal availability of services, such as relocation of opportunities to work, shop, study, and changes in the times of opening of services to alleviate the time constraints of people unable to make convenient use of them at conventional opening times;

promotion of mobile services bringing a wider range of functions directly to rural consumers and reducing the need to travel to larger centres;

improvements in conventional passenger transport services through increasing routes, improving frequencies, and regulating prices; and

introduction of unconventional passenger transport systems including multipurpose services (e.g.

post-buses) and demand-actuated transport (e.g. dial-a-ride and car-pooling).

Whether or not the many alternatives under these five headings are economically or socially feasible and desirable is, of course, the critical question determining their eventual implementation. But Moseley also explored the institutional arrangements that would need to be made. Three features in particular are suggested by the list of options. The first is the potential conflict between attempts to improve rural accessibility under the last four options (which, *inter alia*, makes the least accessible places more attractive to live in) and the aims and instruments contained in the county's settlement policy, discussed above, which would support the first option. The second is the large number of agencies which would be directly or indirectly concerned or affected in implementing any of the options (Table 9.2). Here we see levers and constraints as opposite sides of the coins. Getting an agency to co-operate (e.g. the regional health authority to provide more mobile services) is an important lever in the pursuit of greater accessibility; failing to get co-operation is a constraint on its achievement. In addition, there is a

complex web of interrelationships among agencies and operators involved in the accessibility question such that a change implemented by one has significant implications for others. The third feature is that the options contain items which would require an immediate and direct outlay of investment (e.g. mobile services, or additional rural bus subsidies) and those which mainly depend on changes in operating conditions such as the introduction of post-buses, or changes in the opening times of services. Difficulties arise in the former category where the investment cuts across the efficiency objectives of public-resource allocation and in the latter category where the principle of *ultra vires* may need to be set aside before an agency with a well-defined statutory role may take on a new role.

Summary

Institutions provide the channels for many kinds of formal and informal actions and organisations. Their content may either facilitate or constrain alternative uses that can be made of resources and landscapes.

There are many issues concerning the structure of institutions which geographers are not well qualified to discuss. For example, their syntax and procedural forms often lie outside the understanding of all but those with a legal training. However, formal institutions do not exist merely for the employment of lawyers. They arise out of need; they concern real resources and relationships; and they require those with skills and interest in these matters to contribute to their formulation and assessment.

Some of these areas of interest have been discussed in earlier chapters: the identification of problems requiring new institutions to bring about improvements (e.g. in the distribution of economic opportunities or the protection of environmental amenities); the recognition of ways in which these, and similar, problems are exacerbated by existing institutions (e.g. the fragmentation of government areas); and the awareness of institutions acting as levers and constraints in the preparation of policies and plans.

Carried to fruition, these interests can also lead to involvement in the creation and working of institutions. Often the passage of new legislation poses problems for those charged with implementing it; clauses defining their powers, procedures and re-

Table 9.2 The impact of different agencies upon the location of people, activities, and personal mobility (Moseley *et al.* 1976)

Impact of upon	Person location	Activity location	Personal mobility
County education dept.	1	2	1
County planning dept.	2	2	0
County social services dept.	0	1	1
County surveyor's dept.	1	0	2
District council	2	2	1
Parish council	1	1	0
British Rail	0	1	2
National Bus Co.	1	1	2
Independent bus firms	0	0	2
Regional health authy.	0	2	1
Regional water authy.	2	1	0
Dept. of Environment	2	2	0
Dept. of Transport	1	1	2
Post Office Corporation	0	1	0
Area Traffic Commissioners	0	0	2
Retailers	1	2	0
Employers	2	2	1

Key 0 = no significant impact of the agency upon the relevant parameter
1 = some impact
2 = considerable impact

sponsibilities have to be interpreted or refined. For example, recent environmental legislation in New South Wales (Hort and Mobbs 1979) calls for environmental impact statements (EIS) where developments are 'likely to have a significant impact on the amenity of a locality'. Now, while many developments would fall quite clearly on one or other side of the line, there are probably just as many which raise doubts. Hence, it can be argued, in such cases there is a prior need for clearer guidelines. The alternative is to let things 'shake out' through planning appeals and court procedures which might eventually result in a series of precedents. In New South Wales, the relevant authority (the Planning and Environment Commission) contracted others to provide discussion papers to help clarify the issues. Quickly it became apparent that at least two classifications of developments needing EIS could be drawn up: one based on the character of development (its size, production processes, toxicity of materials, generation of traffic and noise, etc.), and the other based on location. Doubtless, there will still be problems of interpretation, but this example indicates how practical contributions can be made to the working of a piece of legislation.

Conflict is a normal condition in society. Resources are not unlimited: they can be used in different ways to suit the demands of different people. Their uses are facilitated or restricted by formal and informal rules as well as by conditions in the market place. In so far as geographers have a role to play in ameliorating conflicts, it lies in their contribution to understanding the structure and content of those institutions that have a clear influence upon physical and human landscapes.

10 Conclusion

Applied geography is not a sub-discipline; nor can it ever become one for it lacks an internal corpus of theory and has no clearly definable subject matter. On one hand, it has a dependent relationship with academic geography and, on the other hand, it derives its momentum from substantive issues that arise in the social and physical environment. Yet if it lacks a clear definition, it nevertheless has a valuable role to play in stimulating academic geography and in contributing to practical affairs.

It also contains a *modus operandi* which differs from that of academic geography. The latter, in Popper's terms 'starts with problems and ends with problems' (Schlipp 1974). That is, there is a continuous, internalised, cycle of observation, hypothesis, experimentation (or testing), falsification, renewed observation, and so on, occasionally punctuated by a new paradigm. The cycle in applied geography extends into prescription and, almost by definition, has to engage in dialogue with 'outsiders' – those who manage resources or initiate change. It also contains another meaning of falsification. The reasons why a policy or plan turns out to be unsuccessful may be quite different from the reasons why a hypothesis proves to be invalid. A primary condition for a hypothesis is that it should be internally consistent. A plan or policy needs more than this: it is also required to be externally acceptable.

In retrospect this is why the preceding chapters have emphasised concepts and processes that are not commonplace in academic geography. Several 'key words' can be indicated: wants and ideals, accounts and budgets, uncertainty and risk, choice and conflict, evaluation and assessment. It also explains why there have been occasional forays into other disciplines such as political science and economics. To do otherwise, and stay firmly rooted within geography, would be to run the same risk as those technologists who see the world only in terms of what is technically feasible.

Growing interest in the application of geographical studies has been paralleled by the publication of journals and monographs with the word 'applied' in their title, and by 'applied' courses in universities and colleges. Yet one may suspect that sometimes the word is used loosely: that it expresses little more than an antonym for theory, or that it describes a course that is somehow meant to be linked with the perceived needs of an industry in which students might eventually be employed. If these suspicions are true then we are doing a disservice to our students and to our discipline.

The crux of applied geography is (at the risk of tautology) fundamentally that it is about geography. That is, it deals with human and physical landscapes; their structure and composition, their internal dynamic, the forces that shape them, the choices that exist in the way they can be used. Ultimately, it is about the creation of new geographies.

This does not preclude specialist interests (such as tourism, or housing, or natural resources, or transport) in which there are issues that invite a geographical contribution. However, if these specialisms only involve a one-way transfer (i.e. existing knowledge applied to external problems) then applied geography and its parent discipline will gain nothing. If applied geography is to be more than a static organ through which the current state of geographical knowledge passes, and if it is to make an active contribution to the evolution of the discipline as a whole, then, through praxis, it must contribute to theory and methodology which enable a better understanding of human and physical landscapes and environments.

When emphasis is put on prescription (as it has been throughout this book) then it is necessary to ask what the implications are for the teaching of the subject.

Firstly, to use a medical analogy, one cannot have effective prescription without diagnosis and treatment. (The alternative is doctors with magnificent bed-side manners, but nothing else.) For us, the equivalent of diagnosis lies in a knowledge and understanding of the substantive content of geography – the elements of human and physical systems, and their relationships. Treatment in our terms, means being able to propose actions to meet problems, knowing what resources are available and how they can be used, and what the direct and indirect effects will be. In short, to use terms used earlier, it requires a knowledge of both the 'system to be controlled' and the 'controlling system'.

The substantive content of a discipline usually looks after itself. Within their broadly unifying themes (man–land, spatial interaction, environments) there is a general conjunction at any given time between what geographers do and what concerns society as a whole. It would be difficult to see how they could do otherwise. Hence the long continuity of interest found in most branches of the discipline. The same questions, perhaps in slightly different guises, have a habit or reasserting themselves.

More subject to change are the ways those questions are posed and the methods by which answers are approached. In emphasising the prescriptive nature of applied geography we are, in effect, discussing another way of asking questions which, while not novel, is not very familiar. We are doing more than seeking to describe and explain: we are also seeking assessment of feasibility and (much more important) of worth, and models of management.

In this context the recent attention given to general systems theory (Chapman 1977; Bennett and Chorley 1978) may eventually prove useful, although at present it contains a challenge which, in some circumstances, appears to be insurmountable. This is to design systems which correspond accurately to complex real conditions and whose results generate sufficient confidence to be used in prescription. While encouraging experimentation we should also note the strong case put forward by Kennedy (1979) that it is premature, given our present state of knowledge of man and environment, to abandon traditional methods of enquiry in favour of one which calls 'for an elimination of the historical or, indeed, the configuration [i.e. relating to and/or determined by

unique conditions of time and space] from our studies; for an end to the search for increased understanding; for an abandonment of scientific method as it is generally understood; and for a wholesale adoption of techniques of mathematical description and prediction derived from those employed in information theory and control engineering' (p. 558, my parentheses).

Kennedy's reference to 'scientific method' is fully applicable to applied geography. Notwithstanding the comments on the *modus operandi* of academic and applied geography at the start of this chapter, the *rational* acceptance of a prescription requires confidence in its foundations and replicability.

Moreover, I would strongly assert that it is fully consistent with a commitment to scientific method, in so far as it can be pursued in a social or environmental science, to require of ourselves and our students an awareness of issues and their political salience and an empathy for the objects of our studies.

Empathy is not necessarily about 'taking sides', although that may follow from it. Rather it means putting oneself into the role of others and asking: given their motives and resources what would I have done or wish to do in response to this particular problem? And how might my actions change if my motives and resources were different? Putting oneself in the role of others is never easy, even when one is an otherwise dispassionate observer, but that does not detract from its importance in understanding behaviour and its two-way relationships with landscapes. (The only condition in which it would be unnecessary would be when people were identical in every respect.) Empathy is not something that can be taught, but it can be encouraged and guided through role-playing and field work, and also by exposure to issues of the kind in which students may eventually become involved professionally.

Finally, it is unimportant whether or not applied geography achieves the 'formal' status of a sub-discipline. It will continue to function as long as there are challenging problems in the way that landscapes and environments are used. No one can guarantee that it will provide the right answers, but at least it will stand a fair chance of success if it asks the right questions in the right way.

References

Abler, R., Adams, J.S. and Gould, P. (1972) *Spatial Organisation*, Prentice-Hall, London.

Ackerman, A.E. (1962) Public policy issues and the professional geographer, *Annals A.A.G.*, **52**, 292–8.

Adams, J. (1970) Westminister: The fourth London airport? *Area*, **2**, 1–8.

Adams, J.S. (1972) The geography of riots and civil disorders, *Econ. Geog.*, **48**, 24–42.

Albaum, M. (ed.) (1973) *Geography and Contemporary Issues*, Wiley, New York.

Alexander, I. (1978) The planning balance sheet: an appraisal, pp. 46–68, in McMaster, J.C. and Webb, G. R., *Australian Project Evaluation: Selected Readings*, ANZ, Sydney.

Allan, J.A. (1978) Remote sensing in physical geography, *Prog. in Phys. Geog.*, **2**, 36–54.

Allen, K. (1970) The regional multiplier: some problems in estimation, pp. 80–96, in Cullingworth, J.B. and Orr, S. (eds), *Regional and Urban Studies*, Allen and Unwin, London.

Allen, K. and Maclennan, M.C. (1970)*Regional Problems and Policies in Italy and France*, Allen and Unwin, London.

Allison, L. (1975) *Environmental Planning: a Political and Philosophical Analysis*, Allen and Unwin, London.

Almond, G.A. and Powell, G.B. (1966) *Comparative Politics: a developmental approach*, Little, Brown, Boston.

Alonso, W. (1964) *Location and Land Use*, Harvard U.P., Camb., Mass.

Amadeo, D. and Golledge, R. (1975) *An Introduction to Scientific Reasoning in Geography*, Wiley, New York.

Appleton, J. (1975) Landscape evaluation: the theoretical vacuum, *Trans. I.B.G.*, **66**, 120–3.

Archer, B.H. (1976) The anatomy of a multiplier, *Reg. Stud.*, **10**, 71–8.

Archer, B.H. and Owen, C.B. (1971) Towards a tourist regional multiplier, *Reg. Stud.*, **5**, 289–94.

Arrow, K.J. and Lind, R.C. (1970) Uncertainty and the evaluation of public investment decisions, *Amer. Econ. Rev.*, **60**, 364–78.

Bain, J.S. (1973) *Environmental Decay: Economic Causes and Remedies*, Little, Brown, Boston.

Baker, A.R.H. (1979) Settlement pattern evolution and catastrophe theory; a comment, *Trans. I.B.G.*, New Series, **4**, 435–7.

Banham, R. (1969) *The Well-Tempered Environment*, Arch. Press, London.

Barkham, J.P. (1973) Recreational carrying capacity: a problem of perception, *Area*, **5**, 218–22.

Barnett, C.J. (1974) *Committee of Inquiry into Local Government Areas and Administration*, Govt. Printer, Sydney.

Barrett, E.C. and Curtis L.F. (1976) *Introduction to Environmental Remote Sensing*, Chapman and Hall, London.

Batty, M. (1976) *Urban Modelling: Algorithms, Calibrations, Predictions*, University Press, Cambridge.

Benevolo, L. (1971) *The Origins of Modern Town Planning*, M.I.T. Press, Camb., Mass.

Bennett, R.J. and Chorley, R.J. (1978) *Environmental Systems*, Methuen, London.

Berentsen, W.H. (1978) Austrian regional development policy, *Econ. Geog.*, **54**, 115–34.

Bernstein, S.J. and Mellon, W.G. (eds) (1978) *Selected Readings in Quantitative Urban Analysis*, Pergamon, Oxford.

Berry, B.J.L. (1970) The geography of the United States in the year 2000, *Trans. I.B.G.*, **51**, 21–53.

Berry, B.J.L. (1971) Monitoring trends, forecasting change and evaluating goal achievement in the urban environment, pp. 93–120, in Chisholm, M., Frey, A.E. and Haggett, P., op. cit.

Berry, B.J.L. (1972) More on relevance and policy analysis, *Area*, **4**, 77–80.

Berry, B.J.L. (1980) Inner city futures: an American dilemma revisited, *Trans. I.B.G.*, New Series, **5**, 1–28.

Biderman, A. (1966) Social indicators and goals, 68–153, in Bauer, R.A. (ed.), *Social Indicators*, M.I.T. Press, Camb., Mass.

Birch, W. (1977) On excellence and problem solving in geography, *Trans. I.B.G.*, New Series, **2**, 417–29.

Black, J. (1977) *Public Inconvenience: access and travel in seven Sydney suburbs*, A.N.U. Press, Canberra.

Blacksell, M. and Gilg, A.W. (1975) Landscape evaluation in practice – the case of southeast Devon, *Trans. I.B.G.*, **66**, 135–40.

Blaikie, P.M. (1975) *Family Planning in India: diffusion and policy*, Edward Arnold, London.

Blaikie, P.M. (1978) The theory of the spatial diffusion of innovations: a spacious cul-de-sac, *Prog. in Human Geog.*, **2**, 268–95.

Blainey, G. (1966)*The Tyranny of Distance*, Sun Books, Melbourne.

Blondel, J. (1969) *An Introduction to Comparative Government*, Weidenfeld and Nicholson, London.

Blowers, A.T. (1974) Relevance, research and the political process, *Area*, **6**, 32–6.

Blue Mountains City Council (1974) *Blue Mountains Draft Strategy Plan*, Urban Systems Corp., Sydney.

Blumenfeld, H. (1955) The economic base of the metropolis, *J. Amer. Inst. Planners*, **21**, 114–32.

Boal, F.W. (1974) Territoriality in Belfast, pp. 191–212 in Bell, C. and Newby, H., *The Sociology of Community*, Cass, London.

Boddy, M.J. (1976) The structure of mortgage finance: building societies and the British social formation, *Trans. I.B.G.*, New Series, **1**, 58–71.

Boudeville, J.R. (1966) *Problems of Regional Economic Planning*, University Press, Edinburgh.

Boudeville, J.R. (1972) *Aménagement du Territoire et Polarisation*, Genin, Paris.

Boulding, K.E. (1962) *Conflict and Defence: a general theory*, Harper, New York.

Bourne, L.S. (1968) Market, location and site selections in apartment construction, *Canadian Geog.*, **12**, 211–226.

Bourne, L.S. (1974) Forecasting urban systems: research design, alternative methodologies and urbanisation trends with Canadian examples, *Reg. Stud.*, **8**, 197–210.

Bourne, L.S. (1975) *Urban Systems: strategies for regulation*, Clarendon Press, Oxford.

Brewis, T.N. (1969) *Regional Economic Policy in Canada*, Macmillan, Toronto.

Brinkman, R. and Young, A. (1976) *A Framework for Land Evaluation*, F.A.O., Rome.

Broadbent, A. (1977) *Planning and Profit in the Urban Economy*, Methuen, London.

Bronfenbrenner, M. (ed.) (1969) *Is the Business Cycle Obsolete?* Wiley, New York.

Brooks, E. (1974) Government decision-making. *Trans. I.B.G.*, **63**, 29–40.

Brooks, E. (1976) The geographer as politician, pp. 60–6 in Coppock, J.T. and Sewell, W.R.D., op. cit.

Brown, A.J. (1972) *The Framework of Regional Economics in the United Kingdom*, University Press, Cambridge.

Brown, A.J. and Burroughs, E.M. (1977) *Regional Economic Problems*, Allen and Unwin, London.

Brunn, S.D. (1974) *Geography and Politics in America*, Harper and Row, New York.

Brunsden, D. and Thornes, J.B. (1979) Landscape sensitivity and change, *Trans. I.B.G.*, New Series, **4**, 463–82.

Buchanan, R.O. (1968) The man and his work, pp. 1–12, in Institute of British Geographers, op. cit.

Buck, T.W. (1970) Shift and share analysis – a guide to regional policy?, *Reg. Stud.*, **4**, 445–50.

Burnham, C.P. and McRae, S.G. (1974) Land judging, *Area*, **6**, 107–11.

Burton, I. (1963) The quantitative revolution and theoretical geography, *Canadian Geog.*, **7**, 151–62.

Burton, I. and Kates, R.W. (1964) The perception of natural hazards in resource management, *Nat. Resource J.*, **3**, 412–41.

Burton, I., Kates, R.W., White, G.F. (1978) *The Environment as Hazard*, Oxford University Press, New York.

Cameron, G.C. (1974) *Regional Economic Development: The Federal Role*, Johns Hopkins, Baltimore.

Cameron, G.C. (1974) Regional economic policy in the United Kingdom, pp. 1–41, in Sant, M.E.C. (1974b), op. cit.

Camina, M.M. (1974) *Local Authorities and the Attraction of Industry*, Pergamon, Oxford.

Canter, L.W. (1977) *Environmental Impact Assessment*, McGraw-Hill, New York.

Cantilli, E.J. (1978) Policy and policy models in transportation, pp. 97–110, in Bernstein, S.J. and Mellon, W.G., op. cit.

Carlisle, E. (1972) The conceptual structure of social indicators, pp. 23–32, in Shonfield, A. and Shaw, S., op. cit.

Carlstein, T. (1977) Regional or spatial sociology? pp. 489–504, in Kuklinski, A., op. cit.

Carlstein, T. (1981) *Time Resources, Society and Ecology*, Vol. 1, Preindustrial Societies, Allen and Unwin, London.

Cazes, B. (1972) The development of social indicators, pp. 9–22, in Shonfield, A. and Shaw, S., op. cit.

Chadwick, G. (1971) *A Systems View of Planning*, Pergamon, Oxford.

Chandler, R.J. (1977) The application of soil mechanics methods to the study of slopes, pp. 157–82, in Hails, J., op. cit.

Chapman, G.P. (1977) *Human and Environmental Systems: an appraisal*, academic Press, London.

Chisholm, M.D.I. (1967) General systems theory and geography, *Trans. I.B.G.*, **42**, 45–52.

Chisholm, M.D.I. (1976) Academics and government, pp. 67–85, in Coppock, J.T. and Sewell, W.R.D., op. cit.

Chisholm, M.D.I. and Manners, G. (1971) Geographical space: a new dimension of public concern and policy, pp. 1–23, in Chisholm and Manners (eds), *Spatial Policy Problems of the British Economy*, University Press, Cambridge.

Chisholm, M.D.I., Frey, A.E. and Haggett, P. (eds), (1971) *Regional Forecasting*, Butterworth, London.

Chorley, R.J. (1971) Forecasting in the earth sciences, pp. 121–38, in Chisholm, M.D.I., Frey, A.E. and Haggett, P., op. cit.

Christians, C. (ed.) (1968) *Colloque Internationale de Géographie Appliquée*, International Geographical Union, Liège.

Clawson, M. (1975) Economic and social conflicts in land use planning, *Nat. Res. J.*, **15**, 473–89.

Clayton, K.M. (1974) Discussion, pp. 261–2, in Sant (1974a), op. cit.

Coates, B.E., Johnston, R.J. and Knox, P.L. (1977) *Geography and Inequality*, University Press, Oxford.

Collins, L (1975) *An Introduction to Markov Chain Analysis*, Institute of British Geographers, London.

Commonwealth of Australia (1979) *Study Group on Structural Adjustment: Report*, Vol. 1, Canberra.

Cooke, R.U. and Harris, D.R. (1970) Remote sensing of the terrestrial environment – principles and progress, *Trans. I.B.G.*, **50**, 1–23.

Cooke, R.U. and Johnson, J.H. (eds) (1969) *Trends in Geography*, Pergamon, Oxford.

Coppock, J.T. (1976) Geography and public policy: challenge, opportunity and implications, pp. 1–19, in Coppock and Sewell, op. cit.

Coppock, J.T. and Sewell, W.R.D. (1976) *Spatial Dimensions of Public Policy*, Pergamon, Oxford.

Coventry City Council (1971) *Coventry–Solihull–Warwickshire: A Strategy for the Sub-Region*, Coventry.

Cox, K.R. (1972)) *Man, Location and Behaviour*, Wiley, New York.

Cox, K.R. (1973) *Conflict, Power and Politics in the City: a geographic view*, McGraw-Hill, New York.

Cox, K.R. (1974) Territorial organization, optimal scale and conflict, pp. 109–39, in Cox, Reynolds and Rokkan, op. cit.

Cox, K.R. (ed.) (1978) *Urbanization and Conflict in Market Societies*, Methuen, London.

Cox, K.R., Reynolds, D.R. and Rokkan, S. (eds) (1974) *Locational Approaches to Power and Conflict*, Sage, New York.

Crofts, R.S. (1975) The landscape component approach to landscape evaluation, *Trans. I.B.G.*, **66**, 124–9.

Cutt, J.C. (1978) Program budgeting and analytical support systems, pp. 8–21, in McMaster and Webb (1978) op. cit.

Czamanski, S. (1972) *Regional Science Techniques in Practice*, Lexington, Mass.

Czamanski, S. (1973) *Regional and Interregional Social Accounting*, Lexington, Mass.

Dales, J.H. (1968) *Pollution, Property and Prices*, University Press, Toronto.

Darwent, D.F. (1969) Growth poles and growth centres in regional planning – a review, *Env. and Planning*, **1**, 5–32.

Dasgupta, A.K. and Pearce, D.W. (1972) *Cost-Benefit Analysis*, Macmillan, London.

Dawson, J.A. and Doornkamp, J.D. (eds) (1973) *Evaluating the Human Environment*, Edward Arnold, London.

Dee, N., Baker, J., Drobny, N., Duke, K., Whitman, I, and Fahringer, D. (1973) An environmental evaluation system for water resource planning, *Water Resources Research*, **9**, 523–35.

Dempsey, R. (ed.) (1974) *The Politics of Finding Out*, Cheshire, Melbourne.

Dennis, N. (1978) Housing policy areas: criteria and indicators in principle and practice, *Trans. I.B.G.*, New Series, **3**, 2–22.

Department of the Environment (1975) *Strategic Plan for the North West*, HMSO, London.

Department of Urban and Regional Development (1976) The urban and regional budget, pp. 531–66, in McMaster and Webb (1976), op. cit.

Deutsch, K.W. (1970) *Politics and Government*, Houghton Mifflin, Boston.

Deutsch, M. (1973) *The Resolution of Conflict*, Yale U.P., New Haven.

Dickinson, G.C. and Shaw, M.G. (1977) What is land use? *Area*, **9**, 38–42.

Dickinson, R.E. (1969) *The Makers of Modern Geography*, Praeger, New York.

Donnison, D. (1974) Regional policies and regional government, pp. 189–99 in Sant (1974b) op. cit.

Douglas, I. (1973) Water resources, pp. 57–87, in Dawson and Doornkamp, op. cit.

Downs, A. (1957) *An Economic Theory of Democracy*, Harper and Row, New York.

Drever, J. (1967) *A Dictionary of Psychology*, Penguin, London.

Drewett, R. (1971) Land values and urban growth, pp. 335–58, in Chisholm, Frey and Haggett, op. cit.

Duffield, B.S. and Coppock, J.T. (1975) The delineation of recreational landscapes: the role of a computer based information system, *Trans. I.B.G.*, **66**, 141–8.

Duncan, S.S. (1974) Cosmetic planning or social engineering? Improvement grants and improvement areas in Huddersfield, *Area*, **6**, 259–71.

Duncan, S.S. (1976) Research directions in social geography: housing opportunities and constraints, *Trans. I.B.G.*, New Series, **1**, 10–19.

East Anglia Regional Strategy Team (1974) *Strategic Choice for East Anglia*, HMSO, London.

Enzer, S. (1970) *Delphi and Cross Impact Techniques*, I.F.F., New York.

Finer, S.E. (1970) *Comparative Government*, Allen Lane, London.

Fines, K.D. (1968) Landscape evaluation: a research project in East Sussex, *Reg. Stud.*, **2**, 41–55.

Fishbein, M. and Ajzen I. (1975) *Belief, Attitude, Intention and Behaviour*, Addison-Wesley, Reading, Mass.

Flegg, A.T. (1976) Methodological problems in estimating recreational demand functions and evaluating recreational benefits, *Reg. Stud.*, **10**, 353–62.

Forbes, J. (1969) A map analysis of potentially developable land, *Reg. Stud.*, **3**, 179–95.

Form, W.H. (1954) The place of social structure in the determination of land use: some implications for a theory of urban ecology, *Social Forces*, **32**, 317–23.

Forsythe, D. (1972) *U.S. Investment in Scotland*, Praeger, London.

Freeman, T.W. (1966) Discussion, p. 44, in Strida, M. (ed.), *La Géographie Appliquée dans le Monde*, International Geographical Union, Prague.

Friedmann, J. (1966) *Regional Development Policy : a Case Study of Venezuela*, M.I.T. Press, Camb., Mass.

Galbraith, J.K. (1975a) *Economics and the Public Purpose*, Penguin, London.

Galbraith, J.K. (1975b) *Economics, Peace and Laughter*, Penguin, London.

Gilbert, E.W. (1951) Geography and regionalism, pp. 345–71, in Taylor G. (ed.), *Geography in the Twentieth Century*, Methuen, London.

Gillingwater, D. (1975) *Regional Planning and Social Change: a responsive approach*, Saxon House, Farnborough.

Glasson, J. (1974) *An Introduction to Regional Planning*, Hutchinson, London.

Goddard, J.B. (1975) *Office Location in Urban and Regional Development*, University Press, Oxford.

Godlund, S. (1971) Regional and sub-regional population forecasts: current Swedish practice, pp. 309–322, Chisholm, Frey and Haggett, op. cit.

Goodey, B. (1970) Mapping utopia, *Geog. Rev.*, **60**, 15–30.

Gordon, I.R. (1973) The return of regional multipliers: a comment, *Reg. Stud.*, **7**, 257–62.

Gottman, J. (1973) *The Significance of Territory*, University Press of Virginia, Charlottesville.

Gould, P. (1963) Man against his environment: a game-theoretic framework, *Annals A.A.G.*, **53**, 290–7.

Gould, P. and White, R. (1974) *Mental Maps*, Penguin, London.

Gray, F. (1976) Selection and allocation in council housing, *Trans. I.B.G.*, New Series, **1**, 34–46.

Green, D.H. (1977) Industrialists information levels about regional incentives, *Reg. Stud.*, **11**, 7–18.

Greenberg, M.R., Anderson, R. and Page, G.W. (1978)

Environmental Impact Statements, Assoc. of Amer. Geog., Commission on College Geography, Resource Paper 78(3).

Greenhut, M. (1956) *Plant Location in Theory and Practice*, University of N. Carolina Press, Chapel Hill.

Gregory, D. (1978) *Ideology, Science and Human Geography*, Hutchinson, London.

Grigg, D. (1965) The logic of regional systems, *Annals A.A.G.*, 55, 465–91.

Hagerstrand, T. (1967) *Innovation Diffusion as a Spatial Process*, University Press, Chicago.

Hagerstrand, T. (1970) What about people in regional science?, *Papers, R.S.A.*, 24, 7–21.

Hagerstrand, T. (1976) The geographers' contribution to regional policy: the case of Sweden, pp. 243–62, in Coppock and Sewell, op. cit.

Haggett, P. (1965) *Locational Analysis*, Edward Arnold, London.

Haggett, P. (1971) Leads and lags in interregional systems: a study of cyclic fluctuations in the South West economy, pp. 69–95, in Chisholm and Manners, op. cit.

Hails, J. (ed.) (1977) *Applied Geomorphology*, Elsevier, Amsterdam.

Hamblin, R.L. (1975) Social attitudes: magnitude, measurement, and theory, pp. 61–120, in Blalock, H.M. (ed.), *Measurement in the Social Sciences*, Macmillan, London.

Hamilton, F.E.I. (ed.) (1974) *Spatial Perspectives on Industrial Organisation and Decision-making*, Wiley, London.

Hamnett, C. (1973) Improvement grants as an indicator of gentrification in Inner London. *Area*, 5, 252–61.

Hardin, G. (1968) The tragedy of the commons, *Science*, 162, 1243–8.

Hare, F.K. (1976) Geography and public policy in Canada, pp. 42–9, in Coppock and Sewell, op. cit.

Hartshorne, R. (1959) *Perspective on the Nature of Geography*, Murray, London.

Harvey, D. (1973) *Social Justice and The City*, Edward Arnold, London.

Harvey, D. (1974) What kind of geography for what kind of public policy, *Trans. I.B.G.*, 63, 18–24.

Haynes, R.M. and Bentham, C.G. (1979) Accessibility and the use of hospitals in rural areas, *Area*, 11, 186–91.

Herbert, D. (1972) *Human Geography: a Social Perspective*, David and Charles, Newton Abbott.

Hermansen, T. (1969) Information systems for regional development control, *Papers R.S.A.*, 22, 107–39.

Herschell, W. (1801) Observations tending to investigate the nature of the sun, *Phil. Trans.*, 91, 265–318.

Hicks, U.R. (1974) *The Large City: a world problem*, Macmillan, London.

Hill, M. (1968) A goal-achievement matrix for evaluating alternative plans, *J.A.I.P.*, 34, 19–29.

Hill, M. (1973) *Planning for Multiple Objectives*, Reg. Sci. Res. Inst., Philadelphia.

Hocking, D.M. (1971) *Some Economic Effects of Australia's Two-Airline Policy*, University Press, Melbourne.

Hohnen, S.A. (1976) The potential of the Pilbara as a region, pp. 87–101, in Linge, G.J.R. (ed.), *Restructuring Employment Opportunities in Australia*, A.N.U. Press, Canberra.

Holdgate, M.W. and White, G.F. (1977) *Environmental Issues*, Scope Report 10, Wiley, London.

Hollis, M. and Nell, E. (1975) *Rational Economic Man : a philosophical critique of neoclassical economics*, University Press, Cambridge.

Honey, R. (1976) Conflicting problems in the political organisation of space, *Annals Reg. Sci.*, 10, 45–60.

Hoover, E.M. (1954) Some institutional factors in business investment decisions, *A.E.R.*, 44, 201–13.

Hort, L. and Mobbs, M. (1979) *Outline of N.S.W. Environmental and Planning Law*, Butterworths, Sydney.

Hotelling, H. (1929) Stability in competition, *E.J.*, 39, 41–57.

House, J.W. (1966) Questionnaire No. 2 on Applied geography in Great Britain, pp. 177–94, in Strida, M., *La Géographie Appliquée dans le Monde*, International Geographical Union, Prague.

Hughes, J.T. and Kozlowski, J. (1968) Threshold analysis – an economic tool for town and regional planning, *Urban Studies*, 5, 132–43.

Institute of British Geographers (1968) *Land Use and Resources: Studies in Applied Geography. A Memorial Volume to Sir Dudley Stamp*, I.B.G., London.

Jacobs, J. (1962) *The Death and Life of Great American Cities*, Cape, London.

Jantsch, E. (1967) *Technological Forecasting in Perspective*, O.E.C.D., Paris.

Johnston, R.J. (1968) Choice in classification: the subjectivity of objective methods, *Annals A.A.G.*, 58, 575–89.

Kahn, H. and Wiener, A.J. (1967) *The Year 2000*, Macmillan, London.

Kamrany, N.M. and Christakis, A.N. (1970) Social indicators in perspective, *Socio-Econ. Planning Sci.*, 4, 207–16.

Kasperson, R.E. and Minghi, J.V. (eds) (1970) *The Structure of Political Geography*, University Press, London.

Kast, F.E. and Rozensweig, T.E. (1972) The modern view: a systems approach, pp. 14–28, in Beishon, J. and Peters, G. (eds), *Systems Behaviour*, Harper and Row, London.

Kates, R.W. (1971) Natural hazard in human ecological perspective: hypotheses and models, *Econ. Geog.*, 47, 438–51.

Kates, R.W. (1978) *Risk Assessment of Environmental Hazard*, Scope Report 8, Wiley, New York.

Keeble, D.E. (1967) Models of economic development, pp. 243–302, in Chorley, R.J. and Haggett, P. (eds), *Models in Geography*, Methuen, London.

Kennedy, B.A. (1979) A naughty world, *Trans. I.B.G.*, 4, 550–8.

King, L.J. (1976) Alternatives to positive economic geography, *Annals A.A.G.*, 66, 293–308.

King, L.J. and Clark, G.L. (1978) Government policy and regional development, *Prog. in Human Geog.*, 2, 1–16.

Kneese, A. (1977) *Economics and the Environment*, Penguin, London.

Knox, P.L. (1975) *Social Well-Being: A Spatial Perspective*, University Press, Oxford.

Kozlowski, J. and Hughes, J.T. (1972) *Threshold Analysis*, Architectural Press, London.

Krutilla, J.V. and Fisher, A.C. (1975) *The Economics of Natural Environments*, Johns Hopkins, Baltimore.

Kuklinski, A. (ed.) (1977) *Social Issues in Regional Policy and Regional Planning*, Monton, The Hague.

Kupper, U.I. (1973) The contribution of German geographers to local boundary reforms, *Area*, 5, 172–76.

Lavery, P. (ed.) (1974) *Recreational Geography*, David and Charles, Newton Abbott.

Layard, R. (ed.) (1972) *Cost-Benefit Analysis*, Penguin, London.

Leach, B. (1974) Race, problems and geography, *Trans. I.B.G.*, 63, 41–7.

Lee, C. (1973) *Models in Planning*, Pergamon, Oxford.

Lee, D. (1973) Requiem for large-scale models, *J.A.I.P.*, 39 (3), 163–79

Lever, W.F. (1972) The intra-urban movement of manufacturing: a Markov approach, *Trans. I.B.G.*, 56, 21–38.

Lever, W.F. (1974) Manufacturing linkages and the search for suppliers and markets, pp. 309–33, in Hamilton, op. cit.

Ley, D. (1977) Social geography and the taken-for-granted world, *Trans. I.B.G.*, 2, 498–512.

Ley, D. and Cybriwsky, R. (1974) Urban graffiti as territorial markers, *Annals A.A.G.*, 64, 491–505.

Ley, D. and Anderson, G. (1975) The Delphi technique in urban forecasting, *Reg. Stud.*, 9, 243–9.

Lichfield, N. (1969) Cost-benefit analysis in urban expansion: a case study; Peterborough, *Reg. Stud.*, 3, 123–55.

Lichfield, N. (1970) Evaluation methodology of urban and regional plans: a review, *Reg. Stud.*, 4, 151–65.

Lichfield, N. and Chapman, H. (1968) Cost-benefit analysis and road proposals for a shopping centre – a case study: Edgware, *J. Trans. Econ. Policy*, 2, 280–320.

Lichfield, N., Kettle, P. and Whitbread, M. (1975) *Evaluation in the Planning Process*, Pergamon, Oxford.

Liddle, M.J. (1976) An approach to objective collection and analysis of data for comparison of landscape character, *Reg. Stud.*, 10, 173–81.

Lindblom, C.E. (1959) The science of muddling through, *Public Admin. Rev.*, 19, 79–88.

Linge, G.J.R. (1967) Governments and the location of industry in Australia, *Econ. Geog.*, 43, 43–63.

Linton, D. (1968) The assessment of scenery as a natural resource, *Scot. Geog. Mag.*, 84, 219–38.

Logan, A. (1979) Recent directions of regional policy in Australia, *Reg. Stud.*, 13, 153–60.

Logan, M., Maher, C.A., McKay, J. and Humphreys, J.S. (1975) *Urban and Regional Australia*, Sorrett, Melbourne.

Losch, A. (1954) *The Economics of Location*, Yale U.P., New Haven.

Lowry, I.S. (1964) *A Model of Metropolis*, Rand Corp., Santa Monica.

Luke, R.H. and McArthur, A.G. (1978) *Bushfires in Australia*, Australian Government, Canberra.

Lukes, S. (1974) *Power: a radical view*, Macmillan, London.

Luttrell, W. (1962) *Factory Location and Industrial Movement*, NIESR, London.

Mabbutt, J.A. (1978) The impact of desertification as revealed by mapping, *Environmental Conservation*, 5, 45–56.

Mabogunje, A.L. (1976) The population census in Nigeria, pp. 207–26, in Coppock and Sewell, op. cit.

McCalden, G. (1973) Design considerations for urban and

regional planning information systems in the Australian context, *URPIS 1*, 24.01–24.35.

McCrone, G. (1969) *Regional Policy in Britain*, Allen and Unwin, London.

McHarg, I.L. (1969) *Design with Nature*, Doubleday, New York.

McKay, D.H. and Cox, A.W. (1979) *The Politics of Urban Change*, Croom Helm, London.

McLoughlin, J.B. (1969) *Urban and Regional Planning*, Faber, London.

McMaster, J.C. and Webb, G.R. (eds) (1976) *Australian Urban Economics*, ANZ, Melbourne.

McMaster, J.C. and Webb, G.R. (eds) (1978) *Australian Projecte Evaluation: Selected Readings*, ANZ, Melbourne.

Malisz, B. (1969) Implications of threshold theory for urban and regional planning, *J. Town Planning Inst.*, 55, 108–10.

Mansfield, N.W. (1971) The estimation of benefits from recreation sites and the provision of newer recreation facility, *Reg. Stud.*, 5, 55–69.

Marris, R. (1958) *Economic Arithmetic*, Macmillan, London.

Marris, R. (ed.) (1974) *The Corporate Society*, Macmillan, London.

Massam, B. (1975) *Location and Space in Social Administration*, Edward Arnold, London.

Massey, D.B. (1973) The basic service category in planning, *Reg. Stud.*, 7, 1–15.

Massey, D.B. (1979) In what sense a regional problem?, *Reg. Stud.*, 13, 233–43.

Mason, B.J. (1970) Future developments in applied climatology: an outlook to the year 2000, *Q. Jl. R. Met Soc.*, 96, 349–68.

Mattila, J.M. and Thompson, W.R. (1955) Measurement of the economic base of the metropolis, *Land Econ.*, 31, 215–28.

Mesthene, E. (1974) On the ideal-real gap, pp. 1–18, in Marris (1974), op. cit.

Miles, I. (1975) *The Poverty of Prediction*, Saxon House, Farnborough.

Mishan, E.J. (1967) *The Costs of Economic Growth*, Penguin, London.

Mishan, E.J. (1972) *Elements of Cost-Benefit Analysis*, Allen and Unwin, London.

Mitchell, B.R. and Deane, P. (1962) *Abstract of British Historical Statistics*, University Press, Cambridge.

Mitchell, C.W. (1951) *What Happens During Business Cycles*, NBER, New York.

Montefiore, A.C. and Williams, W.W. (1955) Determinism and possibilism, *Geog. Studies*, 2, 1–11.

Moore, B. and Rhodes, J. (1973) Evaluating the effects of Britain's regional economic policy, *E.J.*, 83, 87–110.

Moore, B. and Rhodes, J. (1974) The effects of regional economic policy in the United Kingdom, pp. 43–69, in Sant (1974b), op. cit.

Morgan, W.T.W. (1975) *East Africa*, Longman, London.

Morrill, R.L. (1973) Ideal and reality in reapportionment, *Annals A.A.G.*, 63, 463–77.

Morrill, R.L. (1976) The geographic imagination and political redistricting, pp. 227–42, in Coppock and Sewell, op. cit.

Morrison, W.I. (1974) *Simulating the Urban Economy: ex-*

periments with input-output techniques, Pion, London.

Moseley, M.J. (1973a) Growth centres – a shibboleth? *Area*, 5, 143–50.

Moseley, M.J. (1973b) The impact of growth centres in rural regions – II. An analysis of spatial 'flows' in East Anglia, *Reg. Stud.*, **7**, 77–94.

Moseley, M.J. (1974) *Growth Centres in Spatial Planning*, Pergamon, Oxford.

Moseley, M.J. (1979) *Accessibility: The Rural Challenge*, Methuen, London.

Moseley, M.J., Harmon, R.G., Coles, O.B. and Spencer, M.B. (1976)) *Rural Transport and Accessibility*, University of East Anglia, Norwich.

Moss, B. (1976) Ecological considerations in the preparation of environmental impact statements, pp. 82–90, in O'Riordan and Hey, op. cit.

Munn, R.E. (1975) *Environmental Impact Assessment: principles and procedures*, Scope Report 5, Toronto.

Musgrave, R.A. and Musgrave, P.B. (1976) *Public Finance in Theory and Practice*, McGraw-Hill, New York.

Myrdal, G. (1957) *Economic Theory and Underdeveloped Regions*, Duckworth, London.

Nash, K.R. (1978) The Sydney City Council land information system: an instrument for urban planning, *URPIS* 6, 1–18.

Nash, P. (1968) Applied geography and ekistic experience, pp. 53–62, in Christians, op. cit.

Newby, H., Bell, C.R., Rose, D., and Saunders, P. (1977) *Property, Paternalism and Power*, Hutchinson, London.

Newton, T. (1972) *Cost-Benefit Analysis in Administration*, Allen and Unwin, London.

Norfolk County Council (1977) *Norfolk Structure Plan*, Norwich.

North, D.C. (1955) Location theory and regional economic growth, *J.P.E.*, **63**, 243–58.

Norton, R.D. and Rees, J. (1979) The product cycle and the spatial distribution of American manufacturing, *Reg. Stud.*, **13**, 141–52.

O'Dell, P.R. (1976) Energy policies in Western Europe and the geography of oil and natural gas, pp. 164–86, in Coppock and Sewell, op. cit.

O'Dell, P.R. (1978) North Sea oil and gas resources: their implications for the location of industry in western Europe, pp. 11–23, in Hamilton, F.E.I., (ed.), *Industrial Change: International Experience and Public Policy*, Longman, London.

Ollier, C. (1977) Terrain classification: methods, applications and principles, pp. 277–316, in Hails, op. cit.

Ollson, G. (1965) *Distance and Human Interaction*, Reg. Sci. Res. Inst., Philadelphia.

Ollson, G. (1978) On the mythology of the negative exponential, *Geografiska Annaler*, **60 B**, 116–23.

O'Riordan, T. (1968) *Multipurpose Use of Water in Broadland*, Unpub., PhD Thesis, University of Cambridge.

O'Riordan, T. (1976) *Environmentalism*, Pion, London.

O'Riordan, T. and Hey, R. (eds) (1976) *Environmental Impact Assessment*, Saxon House, Farnborough.

Pahl, R.E. (1970) *Whose City*?, Longman, London.

Pahl, R.E. (1971) Poverty and the urban system, pp. 126–45, in Chisholm and Manners (eds), op. cit.

Patmore, J. (1973) Recreation, pp. 224–48, in Dawson and Doornkamp, op. cit.

Pearce, D.W. (1976) Measuring the economic impact of environmental change, pp. 142–66, in O'Riordan and Hey, op. cit.

Perloff, H.S., Dunn, E.S., Lampard, E.E. and Muth, R.F. (1960) *Regions, Resources and Economic Growth*, Resources for the Future, Baltimore.

Perroux, F. (1950) Economic space, theory and applications, *Q.J.E.*, **64**, 90–7.

Perry, N.H. (1969) Geography and local government reform, pp. 233–43, in Cooke and Johnson, op. cit.

Porter, P.W. and Lukerman, F.E. (1976) The geography of utopia, pp. 197–224, in Lowenthal, A. and Bowden, M.J. (eds), *Geographies of the Mind*, Oxford University Press, New York.

Pounds, N. (1972) *Political Geography*, McGraw-Hill, London.

Pred, A. (1966) *The Spatial Dynamics of U.S. Urban Industrial Growth, 1800–1914*, M.I.T. Press, Camb., Mass.

Pred, A. and Tornqvist, G. (1973) *Systems of Cities and Information Flows*, Liber/Gleerup, Lund.

Prescott, J.R.V. (1965) *The Geography of Frontiers and Boundaries*, Chicago.

Prest, A.R. (1960) *Public Finance in Theory and Practice*, Weidenfeld and Nicholson, London.

Ravenstein, E.G. (1885; 1889) The laws of migration, *J.R.S.S.*, **48**, 167–235, and **52**, 241–305.

Relph, E.R. (1976) *Place and Placelessness*, Pion, London.

Richardson, H.W. (1973) *Regional Growth Theory*, Macmillan, London.

Richardson, H.W. (1978) *Regional and Urban Economics*, Penguin, London.

Roberts, M. (1974) *An Introduction to Town Planning Techniques*, Hutchinson, London.

Rodda, J.C. (1970) Rainfall excesses in the United Kingdom, *Trans. I.B.G.*, **49**, 49–60.

Rose, H.M. (1970) The development of an urban subsystem: the case of the Negro ghetto, *Annals A.A.G.*, **60**, 1–17.

Rose, H.M. (1978) The geography of despair, *Annals A.A.G.*, **68**, 453–64.

Rose, R. (1972) The market for policy indicators, pp. 119–41, in Shonfield and Shaw, op. cit.

Rostow, W.W. (1971) *Politics and the Stages of Growth*, University Press, Cambridge.

Rudman, P. (1977) Digital multispectral satellite imagery: its use in regional planning information systems, *URPIS* 5, 3.12–3.31.

Runciman, W.G. (1972) *Relative Deprivation and Social Justice*, Penguin, London.

Saarinen, T.F. (1976) *Environmental Planning: Perception and Behaviour*, Houghton Mifflin, Boston.

Saaty, T.L. (1978) Quantifying conflict resolution with reference to urban and international problems, pp. 257–84, in Bernstein and Mellon, op. cit.

Sackman, H. (1975) *Delphi Critique*, Lexington, Mass.

Sant, M.E.C. (1973) *The Geography of Business Cycles*, L.S.E. Geographical Papers 5, 1973.

Sant, M.E.C. (1974a) *Regional Disparities*, Macmillan, London.

Sant, M.E.C. (1974b) *Regional Policy and Planning for Europe*, Saxon House, Farnborough.

Sant, M.E.C. (1975) *Industrial Movement and Regional De-*

velopment: *The British Case*, Pergamon, Oxford.

Sant, M.E.C. (1976) Inter-regional industrial movement: the case of the non-survivors, pp. 355–70, in Turton, B.J. and Phillips, A.D.M. (eds), *Man, Environment and Economic Change*, Longman, London.

Sant, M.E.C. (1977) Social disparities and regional policy in Britain, pp. 231–72, in Kuklinski, A., op. cit.

Sant, M.E.C. (1978) Issues in employment, pp. 84–105, in Hall, P.G. and Davies, R. (eds), *Issues in Urban Society*, Penguin, London.

Sant, M.E.C. and Moseley, M.J. (1977) *Industrial Development in East Anglia*, Geobooks, Norwich.

Schlipp, P. (1974) *The Philosophy of Karl Popper*, Open Court, La Salle.

Schumm, S. (1977) Applied fluvial geomorphology, pp. 119–56, in Hails, op. cit.

Schumm, S. (1979) Geomorphic thresholds: the concept and its applications, *Trans. I.B.G.*, New Series, 4, 485–515.

Scott, A.J. (1970) Location-allocation systems: a review, *Geog. Analysis*, 2, 95–119.

Scott, A.J. (1971) *An Introduction to Spatial Allocation Analysis*, Commission in College Geography Service Paper No. 9, Association of American Geographers, Washington D.C.

Sewell, W.R.D. and Little, B.R. (1973) Specialists, laymen and the process of environmental appraisal, *Reg. Stud.*, 7, 161–71.

Shonfield, A. and Shaw, J. (eds) (1972) *Social Indicators and Social Policy*, Heinemann, London.

Simmie, J. (1974) *Citizens in Conflict*, Hutchinson, London.

Simmons, I.G. (1974) *The Ecology of Natural Resources*, Edward Arnold, London.

Smith, D.M. (1977) *Human Geography: a welfare approach*, Edward Arnold, London.

Smith, K. (1975) *Principles of Applied Climatology*, McGraw-Hill, London.

Soja, E.W. (1974) A paradigm for the geographical analysis of political systems, pp. 43–71, in Cox, Reynolds and Rokkan, op. cit.

Spry, A.H. (1975) The content of the E.I.S., pp. 20–29, in Australian Conservation Foundation, *The E.I.S. Technique*, Canberra.

Stamp, L.D. (1948) *The Land of Britain: Its Use and Misuse*, Longman, London.

Stamp, L.D. (1957) Geographical agenda: a review of some tasks awaiting geographical attention, *Trans. I.B.G.*, 23, 1–17.

Stamp, L.D. (1960) *Applied Geography*, Penguin, Harmondsworth.

Stamper, R. (1973) *Information in Business and Administrative Systems*, Batsford, London.

Starkie, D.N.M. (1976) The spatial dimensions of pollution policy, pp. 148–63, in Coppock and Sewell, op. cit.

Starkie, D.N.M. and Johnson, D.M. (1975) *The Economic Value of Peace and Quiet*, Saxon House, Farnborough.

Stilwell, F.J.B. (1974) *Australian Urban and Regional Development*, ANZ, Sydney.

Sutton, A J. (1974) *Social Development Regions in the Sydney Metropolitan Region*, Macquarie University, Sydney.

Sutton, A.J. (1977) Problems in developing social information systems, *URPIS 5*, 5.1–5.8.

Taylor, G. (ed.) (1951) *Geography in the Twentieth Century*, Methuen, London.

Taylor, P.J. (1976) An interpretation of the quantification debate in British Geography, *Trans. I.B.G.*, New Series, 1, 129–42.

Taylor, P.J. (1978) Political geography, *Prog. in Human Geog.*, 2, 153–62.

Teitz, M.B. (1968) Toward a theory of urban public facility location, *Papers, R.S.A.*, 21, 35–44.

Thier, J.A. (1979) *Toward a Resolution of Resource Conflict*, Unpub. B.A. Thesis, University of N.S.W.

Thom, R. (1975) *Structural Stability and Morphogenesis*, Reading, Mass.

Thompson, E.J. (1971) Population projections for Greater London, pp. 359–88, in Chisholm, Frey and Haggett, op. cit.

Thorngren, B. (1970) How do contact systems affect regional developments?, *Env. and Planning*, 2, 409–27.

Thrift, N.J. (1977) Time and theory in human geography, *Prog. in Human Geog.*, 1, 65–113, 413–59.

Tiebout, C. (1956) Exports and regional growth, *J.P.E.*, 64, 160–4.

Tornqvist, G (1970) *Contact Systems and Regional Development*, Lund Studies in Geography (B), 35, Lund.

Townroe, P.M. (1971) *Industrial Location Decisions*, Centre for Urban and Regional Studies, Birmingham.

Ullman, E.L. (1954) Amenities as a factor in regional growth, *G.R.*, 44, 119–32.

Ullman, E.L. and Dacey, M.F. (1960) The minimum requirements approach to the urban economic base, *Papers, R.S.A.*, 6, 175–94.

Vernon, R. (1966) International investment and international trade in the product cycle, *Q.J.E.*, 80, 190–207.

Wagstaff, J.M. (1978) A possible interpretation of settlement pattern evolution in terms of 'catastrophe theory', *Trans. I.B.G.*, New Series, 3, 165–78.

Waller, R.A. (1970) Environmental quality: its measurement and control, *Reg. Stud.*, 4, 177–91.

Warner, M.L. and Preston, E.H. (1974) *A Review of Environmental Impact Methodologies*, Battelle, Columbus.

Webber, M.J. (1972) *Impact of Uncertainty on Location*, M.I.T. Press, Camb., Mass.

Weber, A. (1909) *Theory of the Location of Industries*, University Press, Chicago.

Weigert, H.W. (1942) *Generals and Geographers: the twilight of geopolitics*, Oxford University Press, New York.

Weinand, H.C. and Ward, R.G. (1979) Area preferences in Papua New Guinea, *Aust. Geog. Stud.*, 17, 64–775.

Weintraub, E.R. (1975) *Conflict and Cooperation in Economics*, Macmillan, London.

West Midland Regional Study Group (1971) *A Development Strategy for the West Midlands*, Birmingham.

Wheare, K.C. (1963) *Federal Government*, University Press, Oxford.

Whitbread, M. (1977) Problems of measuring the quality of city environments, pp. 92–110, in Wingo and Evans, op. cit.

White, G.R. (ed.) (1974) *Natural Hazards*, Oxford University Press, New York.

Wilcock, A.A. (1975) The geographer before 1800, *Area*, 7, 45–6.

Wildavsky, A.B. (1975) *Budgeting: a comparative theory of budgetary processes*, Little, Brown, Boston.

Wilenski, P. (1978) *Directions of Change: A Review of New South Wales Government Administration*, Govt. of N.S.W., Sydney.

Williams, O.P. (1971) *Metropolitan Political Analysis*, Free Press, New York.

Williams, P.R. (1976) The role of institutions in the inner London housing market: the case of Islington, *Trans. I.B.G.*, New Series, **1**, 72–82.

Williams, P.R. (1978) Building societies and the inner city, *Trans. I.B.G.*, New Series, **3**, 23–34.

Williams, R. (1973) *The City and the Country*, Chatto and Windus, London.

Williamson, J. (1965) Regional inequalities and the process of national development, *Econ. Devt. and Cultural Change*, **13**, 3–45.

Willis, K.G. (1974) *Problems in Migration Analysis*, Saxon House, Farnborough.

Wills, G. (ed.) (1972) *Technological Forecasting*, Penguin, London.

Wilson, A.G. (1974) *Urban and Regional Models in Geography and Planning*, Wiley, London.

Wingo, L. and Evans, A. (1977) *Public Economics and the Quality of Life*, RFF and Johns Hopkins, Baltimore.

Wise, M.J. (1968) Sir Dudley Stamp: his life and times, pp. 261–70, in Institute of British Geographers, op. cit.

Wolpert, J. (1964) The decision process in spatial context, *Annals A.A.G.*, **54**, 1964, 537–558.

Young, A.J. (1973) Rural land evaluation, pp. 5–33, in Dawson and Doornkamp, op. cit.

Author Index

Subject Index